Drone Technology in Architecture, Engineering, and Construction

A Strategic Guide to Unmanned Aerial Vehicle Operation and Implementation

Daniel Tal

Jon Altschuld

Registered Office
John Wiley & Sons, Inc., 111 River Street, Hoboken, NJ 07030, USA

Editorial Office
111 River Street, Hoboken, NJ 07030, USA

For details of our global editorial offices, customer services, and more information about Wiley products visit us at www.wiley.com.

Wiley also publishes its books in a variety of electronic formats and by print-on-demand. Some content that appears in standard print versions of this book may not be available in other formats.

Library of Congress Cataloging-in-Publication Data
Names: Tal, Daniel, 1971- author. | Altschuld, Jon, 1986- author.
Title: Drone technology in architecture, engineering, and construction : a
  strategic guide to unmanned aerial vehicle operation and implementation
  / Daniel Tal, DHM design and DanielTal.com, Denver, CO, USA, Jon
  Altschuld, Chinook Landscape Architecture, Centennial, CO, USA.
Description: Hoboken, NJ, USA : Wiley, 2020. | Includes bibliographical
  references and index.
Identifiers: LCCN 2020020196 (print) | LCCN 2020020197 (ebook) | ISBN
  9781119545880 (paperback) | ISBN 9781119545897 (adobe pdf) | ISBN
  9781119545903 (epub)
Subjects: LCSH: Aerial photography in geomorphology. | Aerial photography
  in municipal engineering. | Drone aircraft in remote sensing. |
  Photogrammetry in architecture. | Micro air vehicles--Industrial
  applications. | Building sites--Location.
Classification: LCC TA593 .T288 2020 (print) | LCC TA593 (ebook) | DDC
  620.0028/4--dc23
LC record available at https://lccn.loc.gov/2020020196
LC ebook record available at https://lccn.loc.gov/2020020197

Cover Design: Wiley
Cover Image: Jon Altschuld/Chinook Landscape Architecture

Set in 10.5/13pts ChapparalPro Regular by SPi Global, Chennai, India
SKY10033175_020722

Contents

Acknowledgments

Jon Altschuld

This book has been created at a crossroads in my life. Shortly after deciding to go out on my own and found Chinook Landscape Architecture, my longtime friend and colleague Daniel Tal approached me about co-authoring a book on drone use and technology. We had collaborated on projects for many years and had even written a business plan for integrating drones into an existing practice when we worked together at a global engineering firm. The idea to write this book was exciting and intimidating, much like the idea of going out on my own. And, as was the case with starting Chinook, there are many friends, colleagues, and family members who made it a possibility.

Thank you to Daniel for always being a collaborative resource and for co-authoring this book with me.

There were several drone industry professionals who were helpful, and I'd like to give a special thanks to Eran Steiner, founder and CEO of AirData UAV, and Joshua Haga of Pix4D. Eran reviewed and added to sections outlining the importance of maintenance and operations logging, and Joshua was a tremendous help in analyzing photogrammetry data and metrics for particularly complex subjects.

Many of the projects highlighted in this book were collaborative efforts. Howard Hume of Yeh and Associates, Inc. and Kevin Shanks of THK Associates, Inc. were key members of several of these projects.

And finally, a big thank you to my wife Brittany and our entire family for putting up with late nights, weekend work, and the overall level of stress that comes with both starting a company and writing a book.

Daniel Tal

There are many people I would like to acknowledge that led to the creation of this book.

My parents, Nissim and Ruth Tal.

Justin Clark, who might as well be an uncredited author for this book. A partner in work and innovation, our synergy and combined energy moves mountains, crashes drones, fries computers and made this book possible. I've never met or worked with a more talented person who I also call friend.

Jenn Becker, Kaitlin Weber, Christ Fortin, Sam Carpenter and Alex Kistler for their boundless humor, dedications to their craft, and making the days fun even when work is not.

The support of my wife, Jenn, and two kids, Anina and Raya, who are the reasons I do anything in my life, I do it for them.

Ann Christensen, Bill Neumann and Laura Kirk, Karen Current, and last but for sure not least Mark Wilcox for their support and belief in the work we do.

DHM Design who has funded my drone work and training; none of this would be possible without them.

PART

1

Introduction to Drone Practice

How to Use This Book

This chapter will provide an overview and road map of the book's content to allow readers to have a clear understanding of how to approach the book's information.

Drones for Architecture, Engineering, and Construction (AEC)

Welcome to *Drone Technology in Architecture, Engineering, and Construction*. This book will delve into the world of drones and how to implement and achieve professional quality results using a drone in a variety of AEC related industries.

First, it is important to define the word *drone* as it relates to this book. The word *drone* is used throughout this book to describe the various flying devices used to collect video footage, images, and other data. Two more technically correct terms for these devices are *UAVs* (unmanned aerial vehicles) or *UASs* (unmanned aerial systems). All three terms will be used throughout the book.

The primary focus of these chapters is how to implement and use drones as it relates to architecture, engineering, and construction (collectively called AEC) projects.The book is a road map to implement UAVs through process driven steps with an emphasis on AEC projects. As detailed in the parts and chapters below, the book will review specific technology, software, regulations, costs, and practical data, and will provide direction to the implementation and management of drones with an office business cycle and budget.

While detailed, the book is by no means comprehensive, as the technology is changing and updating at a fast pace. However, this book should be more than enough to get any reader started with flying a drone, getting insured/licensed, and collecting basic site and project data (Figure 1.1).

Fig. 1-1: Flying a quadcopter drone for site inventory work in Colorado Springs, Colorado. Pikes Peak can be seen in the background. Source: Jon Altschuld.

The Method Behind This Book

Process is at the heart of this book. With that in mind, readers should go sequentially through the chapters, even if they are already familiar with some of the material. In that case, skim through sections, but it is important to understand all the different levels of logistics involved with using drones for work.

Who Can Use This Book?

This book is intended for anyone interested in using drones, particularly on a professional level. This includes hobbyists, architects, landscape architects, surveyors, engineers, planners, and anyone else working in exterior built and non-built environments.

This book DOES NOT require any prior knowledge or use of a drone. Instead, readers will be provided with the full spectrum of requirements and activities to implement the tools, buy the drone hardware, work with specific software, and understand regulations, permitting, and rules that govern UAVs.

The license and permitting portions of the book focus on the rules in the United States, but the remainder of the content would be appropriate for drone pilots and users anywhere in the world.

The Book Road Map

Part 1: Introduction to Drone Practice serves as a general introduction to drones and provides an overview for the book. In addition, Part 1 provides a big picture approach to how drones fit within a firm, what they can accomplish, and why the technology is easy to implement, if not required. Part 1 includes:

Chapter 1: How to Use This Book provides a quick guide and overview for readers on how to best approach this book. Chapter 2: A Paradigm Shift in Viewing the World discusses the ethics, design implications, and mindset that drone operators and managers should consider when bringing these tools to practice. Chapter 3: Drone Data Visualization as a Full Cycle Tool reviews specific project types that have used drones. These projects span the broad AEC industry and demonstrate how UAVs can be adapted for many types of locations, project goals, and the range of deliverables (Figure 1.2).

Fig. 1-2: Drone view of the I-25 and Cimarron Interchange project part way through construction. Drones can be integrated in pre-design, design, construction, and post-construction phases. Source: Jon Altschuld.

Part 2: Getting off the Ground provides the nuts and bolts of drone ownership, implementation, permitting, and start-up costs. It includes chapters on how to fly a drone, a review of drone hardware, specifics about rules and regulations, and tips to get the best results for specific projects. Chapter 4: Buy In, addresses the challenge of bringing a new technology to a practice. This includes how to answer questions from stakeholders about implementing drones easily, and with minimal investment and full understanding of the benefits. Chapter 5: Getting Started delves into drone hardware, what drones are best to start out with, the types of drone sensors and equipment, computer hardware requirements to process drone data, and photogrammetry and applications and software. Chapter 6: Documentation, Permissions, and License focuses on the drone pilot requirements to legally fly a drone in the US, as well as ethical and insurance considerations. This includes how to obtain permission from clients and locations to fly, access important flight data, and how best to approach liability issues. Chapter 7: Best Practices for Flying delves into the specifics of actual drone flight. This will include how to best start flying a drone to collect videos, still images and specific data, and review autonomous drone flight through the use of apps (Figure 1.3).

Part 3: Acquiring and Working with Drone Data focuses on drone data types and processing. This includes how to acquire, manipulate, and use the various data types produced from drone flights. Chapters provide details on converting photogrammetry data into useful 3D models. Chapter 8: Imagery and Videos provides examples of different types of graphics, videos, still images, and diagrams that can be derived from drone footage. Chapter 9: Photogrammetry provides an overview of collecting and processing data for photogrammetry. This chapter also covers point cloud data, data classification, 3D mesh creation, contour creation, and more. Chapter 10: Working with 3D Models discusses how to bring photogrammetry results into

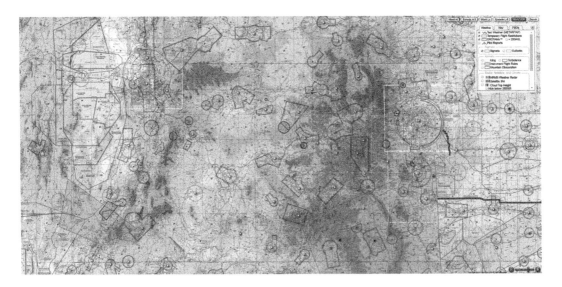

Fig. 1-3: Knowing the rules and regulations for airspace and drone operations is important for legal and safety reasons. This book guides you through the appropriate procedures for commercial drone use in the United States. Source: SkyVector.com.

Fig. 1-4: Drone data, when collected properly, can be used for accurate and precise measurements, existing conditions inventory, and as the base for detailed 3D modeling of proposed design features. Source: Lyons Colorado Drainage and River. Image by DHM Design.

3D visualization software. These chapters include a review of software like Pix4D, Maps Made Easy, SketchUp, Lumion, and other programs that assist in the processing and visualization. Chapter 11: The Future of UAVs briefly looks at other ways drones are being used in professional settings to collect data (Figure 1.4).

About the Authors

Lastly, it is important to introduce the experts behind this book. Each author has a respective website with additional information.

Daniel Tal, ASLA, RLA

Daniel Tal is a practicing landscape architect. He has worked as an urban and site designer since 1998. Daniel graduated from Colorado State University and since developed a specialization in 3D modeling and rendering. He has authored two previous books on 3D modeling: *SketchUp for Site Design* (first edition 2009, second edition 2016) and *Rendering in SketchUp* (2013), both for John Wiley & Sons.

Daniel is a professional presenter and has provided hundreds of presentations and webinars. In 2016, he was the keynote speaker for SketchUp Basecamp in Steamboat, Colorado. In 2017, Daniel was hired by DHM Design in Denver, Colorado, to create a UAV studio and build a practice around drone use for landscape architecture and engineering work.

Daniel develops 3D related applications to assist others in 3D modeling including Place-Maker (www.suplacemaker.com) and UrbanPaint (www.sketchurbanpaint.com), and has a tutorial website for 3D modeling (www.danieltal.com). Daniel's love for drones started when

Daniel Tal.

he met and worked with Jon Altschuld. In collaboration with Justin Clark and Jon Altschuld, Daniel has pursued drone skills and work, including this book.

Jon Altschuld, ASLA, RLA

Jon Altschuld is a registered landscape architect with over 12 years of experience in the design of natural areas, trails and open spaces, ecological restoration, parks, streetscapes, and transportation planning. Throughout his career, he has established himself as a leader in visual communication through emerging technologies such as 3D modeling and drone technology. His visualization work has been featured by the ASLA, SketchUp, Landscape Architecture Magazine, and various other trade publications.

Jon is also a regular presenter and speaker, and has taught numerous software workshops and given professional training on drone data collection, SketchUp, Vue, CAD, Lumion, and Adobe Graphics programs within the planning, architecture, landscape architecture, and construction fields.

Currently, Jon is the principal and founder of Chinook Landscape Architecture, based in Denver, Colorado.

Jon Altschuld.

Important Contributor Justin Clark

We would like to give a special thank you to Justin Clark for his contribution to the knowledge, graphics, and direction for this book. Justin has worked in the urban design and landscape architecture world since 1998. Starting out as a CAD draftsman, he moved his way up to do 3D modeling and IT work, and became a jack-of-all trades with various 3D and CAD software.

Justin is an accomplished drone pilot and master at photogrammetry flights and processing. Daniel and Justin work closely on weekly drone related projects. Justin explores various, innovative ways to fly sites that enhance safety and data acquisition. *Daniel's portions of the book would not have been possible without the work, feedback, and assistance of Justin Clark.* While not an author of this book, his work was key to many of the graphics and flights that are included in this book.

He is an avid outdoors man, relishing in snowboarding, snowmobiles, his truck (Nightmare), and boating (Closer to the Sun).

Justin Clark (right) with Chris Fortin (left).

CHAPTER 2

A Paradigm Shift in Viewing the World

Drones have been in the public lexicon for decades. As military tools, drones have been a focus of research and development for much longer. The earliest unmanned aerial vehicles (UAVs) date back to the 1800s when unmanned balloons with explosives were used for military operations. During WWI and WWII, radio controlled unmanned aircraft were developed for a variety of purposes. What we think of as modern drones were largely developed in warfare in the mid 1980s. During this same time, the industry of hobby radio-controlled aircraft was growing in popularity, and in the 1990s radio-controlled quadcopters and miniature UAVs (drones) began coming to market.

Over the next 15–30 years, public sector drones slowly became cheaper and more widely available. At the same time, several other consumer technologies were quickly advancing. In particular, the progression of drone technology coincided with the digital camera revolution (the first consumer digital camera was released in 1994). The convergence of these two emerging technologies during the early 2000s and into the early 2010s resulted in cheaper and more reliable drones with payloads that were higher quality, more varied, smaller, and lighter. As this happened, drones quickly gained user bases in professional settings ranging from agriculture, to data and image collection, to natural resource monitoring.

Consumers and professionals alike were suddenly exposed to a new way of viewing their surroundings. The low cost of entry, combined with the relatively easy operation, has turned drone use into a standard practice for many industries. It has also opened the door for almost all industries to consider new ways of solving problems (Figure 2.1).

Fig. 2-1: The quality of digital camera sensors plus the accessibility of drones allows us to view our surroundings in new ways. Source: North Fruita Desert Trails Master Plan completed by Chinook Landscape Architecture and THK Associates, Inc.

The Breadth of Drone Applications Is Just Beginning to Be Discovered

Basic Drone Use

In its most basic form, the commercial drone industry is providing a new way to view the natural and man-made world around us. The ability to view a site from above allows us to discover a place in an entirely new way. Drone photography and videography is used all around us, from social media, to real estate agents, to movies and television. For the architecture, engineering, and construction industries, drone imagery provides a major shift both for how we design and how non-designers view our projects. This includes how to effectively communicate design ideas with clients, other members of the design team, and the public. Viewing a site from above also has a powerful impact on the emotional connection one feels with the site (Figure 2.2).

Historically, site design has been presented primarily in plan view. In the last decade, the spread of 3D programs has allowed for the integration of perspective views and interactive 3D graphics of the site itself. However, these are often isolated and lacking in contextual content, and the graphics do a poor job of communicating how the design will fit into its surroundings. Similarly, construction industries have long been limited in how they can effectively track and communicate the status of projects and constantly changing parameters.

Fig. 2-2: Viewing a site from above changes our perspective on topography, connectivity, and context. Source: Conifer High School Sports Fields project completed by ECI Site Construction Management, Inc.

With the widespread use of drones, aerial views of landscapes, plazas, bridges, and buildings are now being viewed by everyone. This gives new significance to contextually sensitive design efforts, and it is also an influential force in shaping how designs will be experienced (Figure 2.3).

Current Breadth of Drone Use

As the drone industry has advanced, so have the applications that drones are used for. In Central America, long-lost ancient cities have been revealed under centuries of rain forest growth by drones equipped with LiDAR sensors. In Colorado, drones are gathering 3D topographic information to create models of rockfaces for geohazard analysis and mitigation along interstate highways. In Utah, ski areas are being mapped by drones for master planning purposes. Across the globe, environmental researchers are using drones to identify and analyze ecological trends, climate change, and wildlife habitat patches/edges. In many areas, drones are flying construction sites to observe and record progress and automatically check built conditions for accuracy/consistency against digital construction drawings. Drones equipped with gas detection equipment are being used to quickly and remotely inspect miles of pipelines for leaks. From inspecting roofs, to real-time assessment of natural disasters, to police surveillance, to automated home security sentries, drones are filling niche after niche (Figure 2.4).

(a)

(b)

Fig. 2-3: Although the snow scene rendering (a) is a great graphic, it has no contextual content and provides no information about what surrounds the proposed building. The drone Photomatch (b), on the other hand, provides the same level of 3D model detail, plus the surrounding conditions and views. Source: Hideway Station project completed by Chinook Landscape Architecture and Viz Graphics.

Fig. 2-4: SketchUp model of an existing rockface along I-70, with proposed rockfall mitigation netting and walls modeled into it. Source: Chinook Landscape Architecture and THK Associates, Inc.

All of these applications focus on utilizing drones to position data collection sensors (LiDAR, cameras, gas detection sensors, etc.) on a mobile, agile, and easy to operate platform. This often results in a cheaper, quicker, and safer alternative to traditional methods. In some cases, such as geohazard analysis in tight canyons or collecting climate data in extremely remote locations, drones are providing a platform to collect data where it would not otherwise be possible. In either case, the applications discussed above are disrupting their respective industries and are forcing those within the industries to re-think the best way to solve problems. And yet, using drones to extend our reach with sensors is just the start.

Uber, Airbus, Altair Aerial, and other companies are scaling drones up in size in order to use them as air taxis. If you think this sounds like science fiction and something that is decades off from happening, the coming years will surprise you. Uber has already selected Dallas, Los Angeles, and Melbourne as pilot cities for their Uber Air project, with test flights beginning in 2020. Although these will be piloted for now, it paves the path for autonomous passenger flights, a trend that regulatory agencies and the public are starting to consider and plan for.

Let's also not forget the emerging use of drones for parcel and package delivery. Although Amazon was initially at the forefront of this effort, Google's Wing project has made major advancements and was granted a certification by the FAA to begin delivering parcels in 2019. Similar to the air taxis, this has a huge influence on how we view our airspace. The legal, ethical, and regulatory implications of these advances are continually evolving.

The Future Breadth of Drone Use

As we consider the mind-boggling ways that consumer and professional drones are currently being utilized, it is logical to wonder what will come next. With so many expanding benefits, we believe that drone technology applications will continue to see advances and new uses for many years and decades, to come. Its unique versatility also allows this technology to adapt and blend with other emerging technological frontiers. Two key technologies that are advancing in parallel to drones are artificial intelligence (AI) and autonomous navigation. As these three technologies continue to develop, their overlap and combined use cases will multiply.

For the architecture, engineering, and construction industries, this will take many forms. Imagine deploying a drone to autonomously fly a site while avoiding crashes, collect topographic data and high-resolution imagery, have it automatically processed into contours and a basemap, and then receive an email when it's done – all from your desk. In terms of laws and regulations, we are not yet at a point where this is possible, but the technology is at the point where this could be done. This could be replicated and applied to construction monitoring, site security, endangered species protection, environmental conditions monitoring, disaster detection, and many other issues.

To take it a step further, image pattern recognition and artificial intelligence could be used to automatically analyze the drone data and extract building footprints, street centerlines, drainage ways, hardscape, etc. Ecopia, an AI driven geospatial company in Toronto, is already doing this as they prepare their service for consumer release. Ecopia is offering the unprecedented technology of taking aerial photos and using image recognition AI to automate the drafting process of computer aided design site plans. They generate survey ready (depending on the source of the data) planimetric drawings that up until this point, could ONLY be done by a human. And this is at a significant reduction in cost which can save companies hundreds of thousands, if not millions, of dollars. The future holds the possibility for true artificial intelligence to make management decisions based on this data. For example, this entire process could be automated to identify potential debris flow catastrophes before they ever occur, in a fraction of the time that it would take humans to arrive at the same conclusions (Figure 2.5).

The most limiting factors for drone technology innovation will likely be regulations, and the invention of other technology. However, regulations vary by city, state, region, and country; worldwide, drone operators will push the envelope of how drones are being used in order to solve problems while staying within regulations. In fact, there is a global incubator occurring around the regulation of automated devices including cars and drones. The country or city that hits upon the right mixture of rules, safety, and productivity will help set the rules and regulations. More than likely it will be the combination of all these different "experiments in rules" that will set the new global paradigm for these devices to work and function in our cities, aiding people. In terms of other technology, as advances and inventions are developed, there will be spikes in the number of ways drones are being used.

Fully automated drone fleets to deliver packages and people, and monitor and gather data is the direction we are heading.

Fig. 2-5: Screenshot of Ecopia data derived from drone photogrammetry information. Using AI Vision software, Ecopia can generate curbs, road, planting areas, trails, and other site data with relative accuracy up to 7 cm. Source: Ecopia Data image produced by Daniel Tal. Image by Daniel Tal, Ambit3D.

The Risks of Drone Technology

Despite all of the benefits, the adoption of drone technology by architecture, engineering, and construction industries, as well as consumer products industries, comes with some manageable risks. These risks vary from understanding what data is being collected, to maintaining a safe air traffic system, to protecting citizen privacy. Regulatory agencies will need to continually adapt and refine rules to avoid risks, but in large part, the responsibility to manage these risks lies with the professional drone operators.

Good Looking Data versus Good Data. At the top of the list of risks is understanding the difference between good looking data and good data. Because of its low cost of entry and ease of use, drone technology is perfectly poised to be widely used, but also widely misunderstood. In particular, drone photogrammetry (discussed in detail in Chapter 9) produces good looking data with very little background knowledge and expertise. However, producing good photogrammetry data requires an understanding of drone technology, geospatial systems, and photogrammetry software/processes. Not understanding how these technologies function together usually results in a 3D site model that looks very impressive, but not very accurate.

Because the 3D site model looks so impressive and detailed, many professionals assume the accuracy matches the detail they are seeing. In reality, the accuracy of the model is hugely dependent on things that professionals are not trained in. These include ground control points, camera angle in relation to the subject, and ground sampling distance (all of these are discussed in detail in Chapter 9). It is not uncommon to see elevation data be as far off as several hundred feet (in relation to projected coordinate systems) prior to adding ground control point data. This is simply a product of the accuracy of the drone's onboard GPS system.

Similarly, it's important to understand how camera angles and flight distance impact data resolution for complex subjects in order to avoid spending several months trying to re-process data into a usable result (Figure 2.6). Unfortunately, the only way to fix the problem is to re-fly the project with a plan that accounts for these factors. One of the authors, Jon Altschuld, has published a detailed online article on the dangers of good looking data versus good data, and the common mistakes people make when looking at drone photogrammetry data, particularly when working with complex and vertical subjects (https://www.linkedin.com/pulse/drones-rockfall-analysis-capturing-complex--vertical-jon-altschuld/).

Using Surveying Tools Does Not a Surveyor Make Building on the last point, it is imperative that any professional drone user understands the accuracy of their data, as well as the legal implications of how that data can be used. A common risk that has become too prevalent in the drone industry is confusing a drone photogrammetric model with a land survey. Currently, professional licensed surveyors are collecting survey data with drones. Although non-surveyor professionals are using the same technology, they lack the rigor or practice of these professionals. Surveyors also use other tools, such as survey grade GPS units, total stations, theodolites, assessor records, and survey benchmarks. Perhaps more importantly, surveying is a licensed profession and attaining a license has both educational and working experience requirements. Surveys stamped by a professional licensed surveyor (PLS) have major legal and financial implications (Figure 2.7).

As such, it's important to understand that non-surveyor collected data should not be considered a survey to be used for construction documentation or operations. If you are not a PLS, do not represent your drone photogrammetry data as a survey. **Period**.

Ethical Drone Practices Questions. Another major risk with the widespread adoption of drone technology is determining what constitutes ethical drone practices. In the United States, drone use is regulated by the Federal Aviation Administration (FAA) and there are published federal rules for commercially operating a drone. However, there are many misconceptions about what is legal, and the quickly evolving field of professional drone deployment is forcing the FAA to adapt and refine these rules. For example, delivery of goods via drone is explicitly prohibited per FAA rules. But because large technology companies are willing to heavily research and invest in this use of drone technology due to consumer demand, the FAA is working with these companies to develop rules and regulations specifically for drone deliveries. As of 2020, both Google's Wing service and Amazon's Prime Air service have been granted FAA approvals for drone delivery services in select areas.

One of the most common misconceptions is that landowners have jurisdiction for the airspace over their property. In fact, the FAA has jurisdiction over all navigable airspace. In spite of this, many state, county, and local agencies have tried to enact drone bans over their property. While they can ban the takeoff and landing on their property, they do not actually have the authority to regulate activities in the airspace over their property.

(a)

(b)

Fig. 2-6: These two images show point clouds of the same subject, but flown from different distances. The first point cloud (a) was flown approximately 30 feet from the subject, while the second (b) was flown approximately 90 feet from the subject. This difference causes a drastic decrease in the number of points (point cloud B has approximately 20% as many points as point cloud A) and the density of points (point cloud B has approximately 9% of the point density compared to point cloud A). Source: Chinook Landscape Architecture.

Fig. 2-7: Even with RTK GPS technology and accurate ground control points, it is not legal, ethical, or a good idea to advertise your products as survey level if you are not a surveyor. Source: North Fruita Desert Trails Master Plan completed by Chinook Landscape Architecture and THK Associates, Inc.

To add to the confusion, many agencies have adopted policies that protect the privacy of citizens. For example, even though they do not have the authority to ban drone flights, they can prohibit the use of drones to spy on, or invade the privacy of, citizens within their properties. On top of this, many municipalities and agencies have created rules internally, but have done a poor job of publishing what these rules are, or how to obtain permission (many have simply tried to "ban drone activities" rather than provide a system for approval). As you can imagine, this has created a confusing patchwork of regulations, laws, and rules for professional drone users.

As with many ethical issues, the best path is often to proceed cautiously, over-communicate, and adhere to rules whenever they can be found. Although flying on or near a person's property does not require you to obtain their approval, it is quick and easy to let them know what you are doing and why. This also goes a long way to promoting the idea of using drone technology for professional purposes, as opposed to being a toy or hobby (Figure 2.8).

Fig. 2-8: Whether it's letting a property owner know you'll be flying nearby, or talking to on-site construction crews, even though the site superintendent knows you'll be flying, over-communication is always a good idea to avoid surprises and confrontations. Source: Chatfield High School Sports Fields project, completed by ECI Site Construction Management, Inc.

Why Use Drones?

When considering all of the benefits and risks of drone technology in professional practice, one of the first questions that often comes up is "why do we need to use drones?" This question is asked by professionals, managers, clients, and academic programs alike. The answer, however, is usually different for each person or company as it largely depends on their current practice and if drones have been integrated into similar fields yet.

For some, drones provide a quicker, cheaper, and more frequent way to collect data compared with traditional methods. For others, drones offer completely new types of data that were previously unattainable, or too expensive to attain. And for some, drones present an entirely new way of addressing an issue or viewing a site condition. This book will go through many of the current trends and uses of drones throughout architecture, engineering, and construction industries. Hopefully, the book will also inspire new ideas and solutions.

Fig. 2-9: Drone photogrammetry point cloud; this data was collected to evaluate sedimentation deposits during construction. Source: Chinook Landscape Architecture and ECI Site Construction Management.

Overall, there is no single answer to why you need to use drones in your practice, but most users find that they solve issues, fill information voids, and force them to think about project sites and their surroundings in a completely new way. The low cost of entry and shallow learning curve for drone technology, combined with its widespread adoption in many industries, makes it difficult to not justify the investment.

The Bottom Line on Drones

Drones are becoming a game-changing technology for many industries, including architecture, engineering, landscape architecture, planning, construction, and other infrastructure related industries. Not only do they provide a new perspective on our surroundings, but they do so at a minimal cost and time investment. As a result, academic programs and companies alike are quickly adopting the technology and uncovering new and unique ways of using it to solve problems and make processes more efficient. These range from gaining new perspectives of existing conditions, to gathering site data, to creating base maps, to monitoring construction progress, to evaluating ecological conditions, to viewing the finished project and gathering post-construction data (Figure 2.9).

We are currently at a point where drones are no longer a hobby or a toy; they are professional tools that have already been widely adopted by firms within the architecture, engineering, and construction industries. Firms and professionals that are not currently using drones in their business are now faced with a decision to either add drones to their practice, sub-contract drone services, or be in a position where they cannot offer a service that is quickly becoming a standard.

CHAPTER 3

Drone Data Visualization as a Full Cycle Tool

Drones are full cycle tools for AEC (architecture, engineering, and construction) industry projects. From start up, development, construction and post-operation, aerial imagery, videos, photogrammetry, and data provide unparalleled information on existing conditions, design review, on-site progress, and use of a project location. This chapter provides an outline of drone services and data and where they integrate with projects. Case studies and project examples will accompany many of the types of products, deliverables, datasets, and uses provided by a drone through various project types and phases (Figure 3.1).

Advantages

Understanding the fundamental advantages of utilizing a drone helps determine the type of project use and deliverables. Five of the fundamental advantages of using drones in AEC projects are discussed below.

Affordable. First, drones have a low cost of entry for a firm to implement, especially compared with other ways to collect similar or even less detailed data. The exact budget for implementing a drone program is reviewed in Chapter 4. Typically, the ROI is pennies on the dollar to purchase, learn to fly, and become compliant with a drone. While it is worth noting that there is a wide range in the price of drones, even lower cost options can deliver high quality datasets (Figure 3.2).

Flexible. Drones are flexible tools. In many cases, a project manager or staff can bring a drone with them to capture data during site visits that need to occur regardless of the drone operations. Similarly, knowledgeable drone pilots can direct others in remote locations with the required flight mission parameters needed to fly a site.

Drones can capture a wide variety of data types. Some drones allow you to switch out camera/sensor types or use multiple cameras/sensors at the same time (DJI Matrice series).

Fig. 3-1: Example of photogrammetry dataset generated from drone data. Bottom left shows the flight path, the top left shows the views captured by the drone and to the right is the generated point cloud created in Pix4D. Source: DHM Design and City of Englewood, CO.

Fig. 3-2: DJI Phantom 4 Pro drone. Source: https://commons.wikimedia.org/wiki/ File:DJI_Phantom_4Pro_04-2017_img3_in_flight.jpg. Licensed under Free Art License.

This means you can affix a LiDAR sensor, multispectral camera, or high-powered optical zoom camera. This allows drones to be unprecedented data gatherers.

Not all drone data is needed for every project and every project will have specific requirements. Maybe it's just high-resolution aerial imagery, some recording footage, or maybe it's more involved and requires photogrammetry, LiDAR, survey, etc. How that data is used determines the data needed. This helps keep a project focused on outcome and budget.

Range and Access. One of the biggest reasons that drones are seeing widespread adoption within the AEC industries is their ability to give users access to difficult to reach areas and

Fig. 3-3: Pix4D desktop software is one example of an application used to process drone photos to create aerials, models, and point clouds. Source: DHM Design.

vantage points. Although there are legal limitations (discussed in detail in Chapter 6), drones are able to capture perspectives of project sites from previously unreachable areas. In many situations, drones are replacing the need for high cost plane or satellite imagery collection, as well as high cost and high risk manual inspection methods (climbing tall structures).

Integration. Drone data is easy to merge into existing project methods and workflows. Analysis is a given by the nature of the drone's vantage point and high-resolution photos and videos. The ability to capture elevation and topographic information provides an almost instant survey. 3D models require less context to generate renderings and simulations by simply merging photos with 3D renders. Drone data processing software, like Pix4D, DroneDeploy, and others, are smart, easy to use tools that extract a variety of data easily shared with others on the project or with other software for further development or analysis (Figure 3.3).

Safety. For some industries, drones represent an ideal way to keep workers and staff safe and at a distance while still being able to gather necessary information. From oil rigs, to power plants, roof inspection, and infrastructure analysis, this ability keeps drone operators out of harm's way and accomplishes the needed tasks. This is undeniably important and a huge benefit.

Project Cycles

Below are examples of how a drone can be used in specific project phases. A drone can be used at any time during this process and is not at a disadvantage of not being used in a prior

Fig. 3-4: Enlargements show detailed objects that are difficult or impossible to see from the ground. This image, taken at the Three Mile Confluence in Glenwood Springs, Colorado, shows part of an irrigation headgate which was not visible from the ground, as well as a rock spiral created by local visitors. Source: Chinook Landscape Architecture and Wright Water Engineers.

phase. However, experience has shown that flying a drone at the first possible part of a project becomes part of the natural workflow: the drone data will assist in design decisions, uncovering hidden or hard to see elements, provide clarity on any conflicts, and use the data throughout the project (Figure 3.4).

Written Proposal

Aerial footage and videos can be used in proposals to better present your team's knowledge of the project site and investment in pursuing the project. Local sites should be quick to fly to get photos and recordings. This also provides an opportunity to quickly and efficiently create initial study and analysis graphics for the proposals (Figure 3.5). Consider including a photogrammetry map that provides a contour file, 3D mesh, and classified 3D point clouds, all of which are excellent as proposal images.

Proposal Interview

For interviews, especially if photogrammetry was captured, a 3D model, photos, and video are valuable analysis tools. They go a long way to separate yourself from the competition and demonstrate that your team is invested in seeing the project succeed. Similarly, CAD, 3D models, or

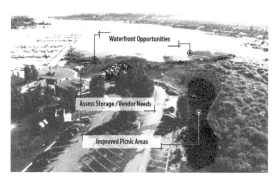

Fig. 3-5: Annotated drone images for a project proposal. The bird's eye images for the Frisco, Colorado, marina were used in PowerPoint and allowed the team to highlight specific project locations for analysis and design considerations. Source: DHM Design.

drawn concepts can be overlaid on photos and aerials of the project to assist with presentation (Figure 3.6).

Project Start Up

For awarded projects, drone images are great for analysis and to share with consultants and clients. This can represent the start of tracking project progress through the use of aerial data. Project teams can overlay design data and continue to update the information as the project progresses.

Concept Design and Design Development

Graphic overlays with 3D models, diagrams, and CAD information, as they are designed, can be integrated with drone footage. Use of photogrammetry 3D models and bird's eye drone images to overlay 3D and 2D concept options are excellent communication tools for presentations and approvals (Figure 3.7).

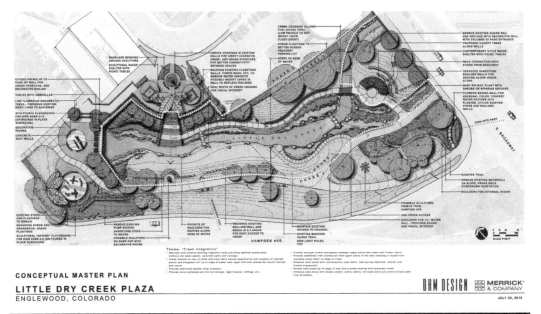

CONCEPTUAL MASTER PLAN

LITTLE DRY CREEK PLAZA
ENGLEWOOD, COLORADO

DHM DESIGN   MERRICK & COMPANY

JULY 20, 2018

Fig. 3-6: The hand drawn site plan (top), is overlaid on the photogrammetric mesh produced from a drone scan of Little Dry Creek in Denver, Colorado. Using 3D modeling software (SketchUp and Lumion) and Photoshop allows for the merger of 3D drone data with 2D hand drawn concepts which is then enhanced with additional 3D modeling trees, people, and environment (bottom). Source: DHM Design and City of Englewood, CO.

Similarly, consultants and the overall design team can easily leverage drone aerial data and 3D models to coordinate pretty much any aspect of design. More context means more team members can be involved and understand a project location; the ability to collect this data early

Fig. 3-7: The existing drone photo (small to the right) is overlaid with a 3D model creating this simple but effective photo match of the new roadway.

on means not having to wait for a complete survey and allows an early start on collaborative design while being sensitive to context. A design team can see the site and design as a whole together allowing for more informed decisions and changes.

Construction Administration

With the ability to re-fly pre-programmed flight paths, typically automated, a construction project can be tracked through time, in almost real-time. This kind of tracking provides accurate quantification of progress, material usage and safety assessments.

Flight data is uploaded to cloud software like Site Aware, Precision Hawk, Trimble Stratus, and others, which then runs comparisons against the previous flights. These intelligent programs allow users to measure slope, perform cut-fill calculations, trace material usage and quantities and measure progress (Figure 3.8). They can even tie into BIM software and Gantt schedule charts. The data and information can be shared between general contractors, subcontractors, design consultants, and clients.

Post-Construction

Drones allow for projects to be monitored after construction and when operations begin: the conditions of the site, location, building, or habitat can be assessed after the start of operations and occupancy; grade can be measured as it starts to settle, geohazards such as landslides or rockfalls can be monitored, infrastructure performance can be tracked and

Fig. 3-8: Drone data in Pix4D and other software can be used to measure quantity, volume, and slope, or provide cut sections and other measurement information. The example shows Lyons, Colorado, earthwork volume calculations. Source: DHM Design.

measured and compared against standards or trends; and design intent compared against as-builts, etc.

Visual Communication

Drone derived products can be used to inform public process, provide outreach, do public meetings, for fundraising and more. Aerial photos or videos, combined with 3D models or analysis diagrams, are an excellent means of communication that are easy to understand in context (Figure 3.9).

Defining Drone Data and Visualization

In order to understand how drones can be used through the stages of a project, it is important to define some of the types of data, graphics, images, diagrams, and information that can be derived from a UAV. While Part 3 goes into greater detail, this section will start by defining and describing these data types. Concepts and data type examples are sprinkled throughout the book.

For the purposes of this book, we will look at data that can be acquired from a drone equipped with a camera. With a camera, there are three main options for collecting data: photos, video, and photogrammetry. Photogrammetry refers to generating 3D models, orthographic aerials, contour data, and other information derived from overlapping photos taken by a drone. Working with photos, videos, and photogrammetry, designers can produce a wide range of deliverables and information beneficial to a project. Many of these deliverables are described below. Where applicable, a real-world example or case study is referenced.

Fig. 3-9: 3D rendered graphic overlays are powerful visuals to show design. A new visitor center was designed and built for the top of Pikes Peak, Colorado. The top image shows the existing condition. The bottom image is the overlay with the proposed design of the new visitor center. Source: DHM Design and RTA/GWWO.

2D Concept Overlay

The merging of a drone bird's eye image with a 2D plan drawing creates a seamless overlay conveying quick conceptual ideas and layouts. See Chapter 8 for more about this process (Figure 3.10).

3D Concept Model Overlay

Merging of a drone bird's eye image with a 3D model of a site concept is done through photo matching the images in photo editing software. See Chapter 8 for more information (Figure 3.11).

3D Mesh

A 3D mesh is a surface created from 3D polygons. Typically, these are either triangular irregular networks (TINs) or quadrangle (quad) meshes. Most photogrammetry software, such as Pix4D and Metashape, will create a 3D mesh. This can then be used in 3D modeling software, including SketchUp and Blender, to model proposed conditions into the existing context.

Fig. 3-10: Diagrams, like the one in the top right, can be overlaid onto a drone bird's eye image of a location as shown for this preliminary master plan upgrade for facilities at Badlands National Park. Source: DHM Design.

Fig. 3-11: 3D models are easily merged with drone photos to create powerful context rich photo matches, like this example of Hideaway Station. Source: Hideaway Station project, completed by Chinook Landscape Architecture and Viz Graphics.

Fig. 3-12: Photogrammetry software can classify data. Trees, buildings, and even road surfaces can be identified and isolated. Source: DHM Design and City of Englewood, CO.

Classified Point Clouds

Software like Pix4D (and its related device app) can generate 3D point cloud data and models from photogrammetry images (see Chapter 9). This data can be classified; meaning trees, vegetation, and man-made structures can be identified and sorted by the software. This classified data can be toggled off from view and removed from 3D models, meshes, and contour data (Figure 3.12).

Construction Analysis over the Internet

Software platforms, often cloud-based, allow for the analysis of a building or a site. One main use is to follow construction progress. Flight data can be used to estimate cut-fill calculations, measure quantities, determine slopes, and compare day-to-day construction progress. These tools are often processed online and are easily accessed by multiple firms, sub-contractors and people (Figure 3.13).

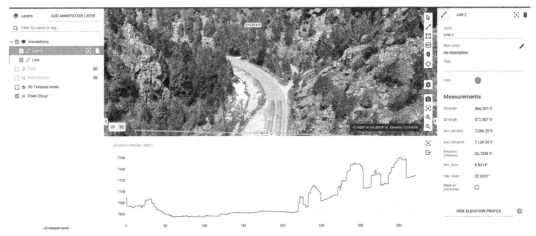

Fig. 3-13: Online cloud-based tools provide access to drone collected data for an entire team. Team members can do volume calculations, cut sections, and generally view the aerials, point cloud, and mesh information. Source: DHM Design.

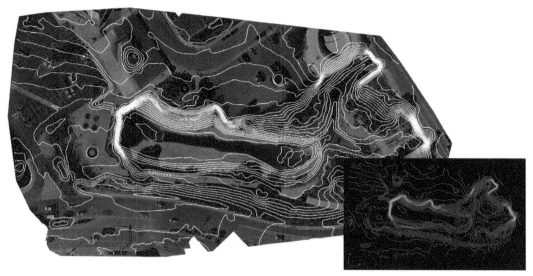

Fig. 3-14: Contour information can be derived during the photogrammetry process for a scanned project site. The contours created for Little Dry Creek are laid over the high-resolution aerial to show how they match up. Source: DHM Design and City of Englewood, CO.

Contour CAD Line Work

The photogrammetry process, using software like Pix4D, can generate contour line work that can be easily used by any software that supports DWG / DXF CAD formats (Figure 3.14).

Ground control survey points are required for accurate contour survey, 3D models, and site data. While survey-grade contours can be produced from proper photogrammetry methods, they are not a substitute for when a survey is legally required unless they have been collected by a surveyor (i.e. for construction drawings). However, this is becoming more common as the technology progresses and more surveyors are using drones to obtain this kind of data.

Drone Photography

Standard cameras, supplemented by lens filters and coupled with apps, can turn even the simplest drone camera into a professional quality flying photo studio with 4K options. Many custom apps are available to enhance photography (Figure 3.15).

Drone Video Footage

Standard high-resolution videos are easy to record and for most drones go up to 4K video resolution.

Fig. 3-15: View of the Badlands National Park captured from the drone used for marketing and project reports. Source: DHM Design.

Using apps like Litchi allows for automated, pre-programmed flight paths providing professional grade smoothness and camera views, not easily achievable with manual flight (Figure 3.16).

Digital Elevation Model (DEM)

A Digital Elevation Model (DEM) is a raster image color coded to show elevation relief and, topography. The color of each pixel represents a specific elevation. DEMs are commonly used in GIS and CAD software, and can be used to create contours and hillshades.

Plant Health

An up-and-coming innovative dataset is the ability to measure vegetation health and spread. Primarily used in agriculture to monitor crop health, it includes the ability to assess if fields are receiving enough or too much water, too much or too little light, exposure to hazardous chemicals, and even monitor pests (Figure 3.17). As software and data become smarter, we should see software able to provide ecological analysis of non-crop, natural ecosystems and even identity plants by species.

High-Resolution Orthorectified Image

True plan high-resolution aerials from drones provide a super high degree of detail, down to the pebbles in some instances. These image files tend to be large (two gigs or more) but printed out, or integrated with project software like BlueBeam, CAD and 3D models, they are excellent coordination tools for an entire project (Figure 3.18).

Fig. 3-16: Videos and multimedia presentations created from drone footage make for excited clients, excellent marketing material, and provide another level of visual data capture useful for a project, like this example of Garden of the Gods in Colorado Springs. Source: DHM Design.

High resolution orthorectified aerials, the product of most photogrammetry processes (see Chapter 9), can be input into software like GIS or CAD and shared among different firms, clients, and contractors. These super high–resolution aerial maps serve as great overlays, can be used to scale and measure for project CAD work, notes, annotations, and layout allowing for greater coordination by teams on a project (Figure 3.19).

Site Quantities and Takeoffs

Pix4D and other photogrammetry software allow for the calculation of areas, volumes and sections from processed drone data. From estimating quantities, to changes in volume, to calculating slope and more, drones are being used on construction sites to monitor and quantify progress.

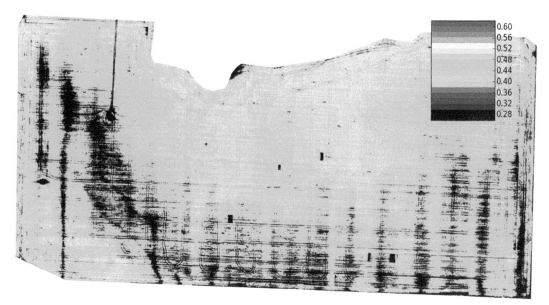

Fig. 3-17: NDVI can capture (using multispectral cameras) moisture levels and measure plant health of crops and other planted locations. Source: https://commons.wikimedia.org/wiki/File:DroneMapper_Processed_NDVI_4cm_GSD. png. Licensed under CC BY-SA 4.0.

Fig. 3-18: High-resolution aerials can capture a great amount of detail.

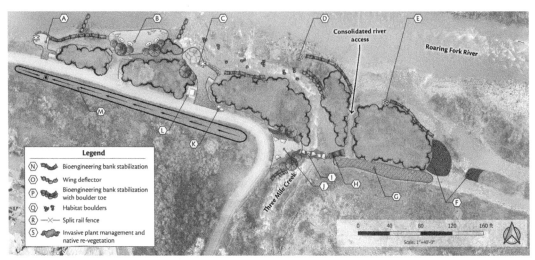

Fig. 3-19: Plan graphics, notes, and design overlays are combined with a high-resolution drone orthorectified image for conceptual design at the Three Mile Creek Confluence in Glenwood Springs, Colorado. Source: Chinook Landscape Architecture and Wright Water Engineers.

Thermal Maps

Thermal maps show real-time heat index for materials like roofs, surfaces, vegetation, and areas. Thermal maps are used to map building heat indexes, rescue missions, and to trace leaks.

Time-Based Site Comparisons

Being able to monitor a construction site over time is saving companies millions of dollars on projects. The relative safety of a drone is allowing construction administration that is easy to repeat and even easier to compare progress (Figure 3.20).

Fig. 3-20: The 88th Ave Open Space and water quality project shown during initial construction (top), continued site grading (middle), and the construction of site amenities and public areas (bottom). Source: DHM Design.

PART
2

Getting off
the Ground

Buy In

Most new technology is met with caution by business owners, principals, and managers. There are specific business questions that need to be addressed and understood before an investment should be made. Drones are no exception. The intent of this chapter is to provide information on how to best pitch the costs, benefits, and return on investment to implement drones into a firm or practice.

Return on Investment (ROI)

When discussing technology, computers, hardware, software, and apps with a business owner, it's sometimes accompanied by a sigh of resignation. For management, those responsible for the firm's budget and finance, investing into a technology like drones, in particular when it's being requested by staff, comes with loads of questions.

Most managers and owners want the answers to the questions listed below. These represent the typical rate of investment versus rate of return.

1. How much does the hardware/software cost?

2. How long will it take to make it profitable? How much can we charge for the service?

3. Which staff should learn to use the technology?

4. How much is training to get up and running?

5. What resources and expertise are available in and out of the office to aid in implementation?

6. Does it fit within our current technology culture and methods/workflows?

Below are systematic answers to the above questions. Starting with a general but accurate cost estimate, each section below further elaborates on the items found in the cost sheet.

Start-Up Cost Estimate

Let's get the hard costs out of the way. Any experienced business owner investing in technology will want to see the estimate first and foremost. Table 4.1 shows the base costs for a drone investment. It's making some assumptions on hourly rates and potential future costs of drones but should serve as a baseline for the overall expenditure. The estimate is broad, but inclusive and detailed. Not all costs will apply to every firm. For example, not everyone will need 3D modeling software. Similarly, Adobe Creative Cloud, AutoCAD, etc. are standard in most firms and might already be accounted for in a budget. The base approximate cost $15,000–16,000 also covers the drone purchase, insurance and labor hours to learn to fly, and become licensed. Annual costs beyond the first investment are less, and are also included below in Table 4.2.

Including photogrammetry or 3D modeling capabilities requires an investment in more software and high-grade computer hardware. Again, for most AEC firms, the software (Photoshop, SketchUp) and hardware are usually already budgeted for and within existing workflows. The specific description related to hardware, software, apps, and training are detailed in other book chapters.

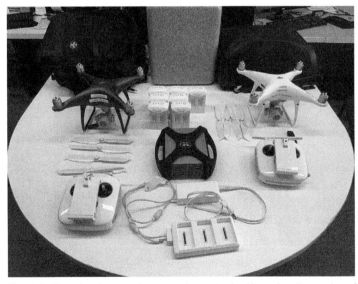

Fig. 4-1: Drone hardware packages can be customized based on the number of drones, batteries, and accessories your company needs. Source: Image by Daniel Tal.

Table 4.1 Base costs for a drone investment.		
Drone Investment Sheet		
All costs in USD		
ITEM		
DRONE	**COST**	**NOTES**
DJI Phantom or Mavic Pro Series drone package	$1,800.00	This is an approximate cost but includes all the base accessories wanted in a drone.
DJI Phantom or Mavic drone		
2× extra batteries		Total of three batteries, one is included with the drone.
Drone backpack		
Controller lanyard		
SD cards for video/photo capture		
Car charger		
Three-stage battery and controller charger		
Total	**$1,800.00**	
LICENSE AND REGISTRATION		
Training material – RemotePilot101.com	$150.00	Best online videos to use.
Employee hours @$100	$4,000.00	40 hours of training time to study for exam and learn to fly the drone.
Drone test exam for Part 107 license	$150.00	Pass the test, get the license.
Drone registration	$5.00	FAA drone registration – required.
Total	**$4,305.00**	
INSURANCE		
Liability insurance	$500.00*	Covers medical or property damage in the case of an accident.
Hull insurance	$150.00*	Covers damage to the actual drone.
Supplemental drone insurance	$150.00	DJI Care replaces two drone losses.
Total	**$800.00**	
*Although drone liability insurance can be purchased separately, many general liability insurance policies will now include drone liability.		

(Continued)

Table 4.1 (Continued)

Drone Investment Sheet		
HARDWARE		
Tablet or smart phone	$300.00	The more current the better. Used to control the drone.
Graphics computer	$2,000.00	Work with photos and videos, create multimedia presentations, process photogrammetry, and create 3D models.
Total	**$2,300.00**	
DEVICE APPS		Apps are for smart devices, software is for computers. Not all costs are applicable to all users.
Litchi app	$35.00	For flying and recording video.
Pix4D flying app	$0.00	Free - for photogrammetry flight planning.
Maps Made Easy Map Pilot for DJI app	$10.00	For photogrammetry flight planning.
Map Pilot for DJI extra features	Ranges from $10–50	Extra features include linear corridor mapping, terrain awareness (from NASA SRTM data), and RAW photo format. Purchase only select features, or unlock all features for $50.
Airmap or SkyVector	$0.00	Free – airspace restrictions check.
UAV forecast	$0.00	Free – weather conditions.
Total	**$45.00+**	
TOTAL COSTS not including software (below)	**$9,250.00**	
SOFTWARE		Apps are for smart devices; software are for computers. Not all costs are applicable to all users.
Pix4D photogrammetry processing software	$3,500.00 annual	Available on month to month subscription as well.
Pix4D training	$4,000	Not a steep learning curve. About 40 hours to learn to use.
Flight logging and maintenance tracking (Airdata UAV)	Ranges free to $15/ month	
Adobe Creative Cloud – photo and video processing	$800.00 annual	Annual cost per user – Photoshop, Premier, Aftereffects, InDesign, etc.
Sketchup 3D modeling software	$600.00 annual	Annual cost.
Lumion 3D rendering software	$3,400.00	3D Rendering software.
Total	**$12,315.00**	
TOTAL COSTS with software	**$21,565.00**	

Annual Expenditures

A quick chart showing maintenance and annual expenditure beyond year one is shown in Table 4.2. Most of these costs are spread out over two years. For example, the drone license exam by the FAA is required every two years and upgrading smart devices can be done every two years. The variables are the drones themselves: does the firm require another drone, more batteries, and/or current drone maintenance? For the estimate below, the full cost of the drone hardware is included to reflect these relative costs. Most firms only require a single drone and the option for more batteries. When new drone models are available, they are worth the upgrade and cost. For DJI Phantom and Mavic series drones, a new line is introduced every two to three years.

Table 4.2 Chart showing maintenance and annual expenditure beyond year one. All costs shown are annual, although many costs would typically be on a two year cycle. For example, a new drone package would be reasonable to purchase every two years; at $1,800, this equates to a $900 annual cost.

Drone Annual Costs		
All costs in USD		
ITEM		
DRONE	COST	NOTES
DJI Phantom or Mavic Pro series drone package	$900.00	This is an approximate cost but includes all the base accessories wanted in a drone. These costs represent the need for another drone, or maintaining the current drone.
DJI Phantom or Mavic Pro series drone		
2× extra batteries		
Drone backpack		
Controller lanyard		
SD cards for video/photo capture		
Car charger		
Three-stage battery and controller charger		
Additional batteries	$180 each	Drone operations will likely require more batteries (for larger/multiple projects), as well as to replace batteries as they age.
Total	$900.00 plus additional batteries	
LICENSE AND REGISTRATION		Costs reflect two year expenditure cycle.
Training material – Remote Pilot101.com	$0.00	Already purchased.
Employee hours @$100	$1,500.00	30 hours of study time every two years.
Drone test exam for Part 107 license	$75.00	Pass the test, get the license.
Total	$1,575.00	

(Continued)

Table 4.2 (Continued)		
Drone Annual Costs		
INSURANCE		
Liability insurance	$500.00*	Covers medical or property damage in the case of an accident.
Hull insurance	$150.00*	
Supplemental drone insurance	$150.00	DJI Care replaces two drone losses.
Total	**$800.00**	
*Although drone liability insurance can be purchased separately, many general liability insurance policies will now include drone liability.		
HARDWARE		
Tablet or smart phone	$150.00	The more current the better. Used to fly the drone.
Graphics computer	$1,000.00	Work with photos and videos, create multimedia presentations, do photogrammetry works, and do 3D modeling.
Total	**$1,150.00**	
DEVICE APPS		One time purchase with start up – no annual costs.
Total	**$0.00**	
TOTAL ANNUAL COSTS	**$4,425.00 plus additional batteries**	

Rate of Return (ROR)

Getting the return on the investment answers the next important questions about budget and financial investment. Below are strategies to monetize a drone service in-house.

Commodity

Drones can be quickly commoditized into project budgets. Designers and engineers with a drone license can easily grab a drone, go on a site visit and, instead of just a camera, use the drone to capture desired data. This can be easily rolled into project scope and fees as part of the rates for project start up. Or just as easily, bump up fees to include drone data capture as part of project site visits and describe the added benefits within the project scope.

To calculate the costs and potential is simple, assuming a local project site. All costs are general and approximate. Context items specific to the project that need to be factored include:

1. Cost to travel to location.
2. Amount of time to fly drone: this can be calculated at roughly 2–4 hours for every 20 acres of site to be flown. These estimates assume the drone team is flying the project to collect photogrammetry information, photos, and videos. If doing automated grid flights, it is also very quick and easy to pre-plan the flights in the office, which will give you a very accurate estimate of flight time.
    i. 20 acres: 2–4 hours
    ii. 40–100 acres: 4–8 hours
    iii. 100+ acres 4-16 and beyond
3. Time to process data: reviewing photos and processing photogrammetry is usually equal to the number of hours flown. So, for every hour of flight time, estimate an hour of processing work; this is the time it takes to set up processing, not how long the software will take to process data. For larger datasets this can vary. Larger projects can vary and this does not include using ground control points or survey information with scans (covered in Chapter 9), which can increase the time required to set up data processing.
4. For remote project sites, adjust the fee based on travel times and travel costs, days on site, etc. Make sure to factor in additional day(s) due to weather conditions.
5. This approach allows for the costs to be included or rolled into the project start up and site visit.

Include in Proposals

It is beneficial to include drone services in a proposal offering the scope of work for project start up, site analysis, photogrammetry, and all the way through construction observation. The advantages for including these services can be sold as:

1. Early collection of key site data like terrain, 3D model, and site analysis assists in project start up and getting a head start on design while using detailed base information obtained through drone collection.
2. Data collection done by qualified pilots who are also designers who understand the project's needs can focus on important and relevant data for a project.
3. Data can be shared across the entire project team allowing for stronger collaboration and better results.
4. Relative low cost of drone scope.
5. Shows the firm has a tech IQ and presence; a cutting-edge advantage.

Similarly, include drones as part of the larger business development strategy. Providing presentations and showing drone work to potential clients will get the word out and help promote a firm's design services and presence (Figure 4.2).

Fig. 4-2: Stunning drone images from past projects make for great marketing materials on future projects. Source: Image by Daniel Tal.

Project Collaboration and Deferred Costs/Cost Savings

A key advantage of drone data and images is the clarity, variety, and quality of the information. Shared among a project team, this information can be used to collaborate on the project. For example, for the 39th Street Greenway project a high-resolution aerial was tied into BlueBeam software allowing the team of consultants (electrical, civil, landscape, utility, etc.) to effectively design to solve issues, develop cost saving strategies, and provide detailed design layouts (Figure 4.3).

All of the collaboration occurs through the cloud (Bentley ProjectWise in this case). At the heart of this collaboration is the drone high-resolution aerial, the survey information, and contours garnished from the drone data.

The cost savings alone pay for the cost of the drone investment and then some.

Having the preliminary site data, contours, aerials, etc. helps with managing the cost of design. Design concepts can be started sooner and tested against site conditions without waiting for a survey and by having detailed reference information to use and check. The information can help in identifying design issues and considerations early on in the process. Consultants can discuss existing conditions and formulate detailed project direction with greater clarity, finding efficiencies in the process. For example, having a preliminary set of grades (2D CAD or 3D) facilitates early ideas and concepts. Detailed high-resolution aerials give more detailed information better than site photos, or free online aerials. The ability to use photogrammetry software to measure volumes, view sections, and overlay other data is unprecedented for most projects in the initial starting phase leading to more informed choices.

Fig. 4-3: Drone and 3D model overlay image of the 39th Ave Greenway concept in Denver, Colorado. Source: DHM Design.

All of these save the project time and, in turn, money. Less time needs to be spent on-site, or revisiting a site; another savings in fees. In some instances, survey costs can be saved (even if these are usually nominal for most projects) if the drone data is collected with a survey accurate level of information in mind.

Intangible Values

Perception of a firm's capabilities is key to clients and important for repeat business. For work done by the authors, the client's perception of designers using drones on projects is one of admiration. Clients get a thrill when viewing their project from above. Clients gain greater confidence in the firm's overall expertise and the perception of the firm's design and technology IQ increases.

The added advantage of a competitive edge against other competitors is relevant; for example, DHM design provided drone services for the City of Thornton, Colorado, as part of a proposal. This helped secure the job for the Thornton Justice Center campus project because of the availability of site information like the aerial and contours, not to mention the photos from above the site. Given how fast drones are being adopted into the design sectors, the industry is quickly approaching a tipping point of needing in-house drone services to maintain a competitive edge.

Another intangible advantage is the ability to start design projects without waiting for a survey. While a professional survey is still required in many instances (see Chapter 9), the combination of high-resolution aerials, contours data, and 3D models produced from

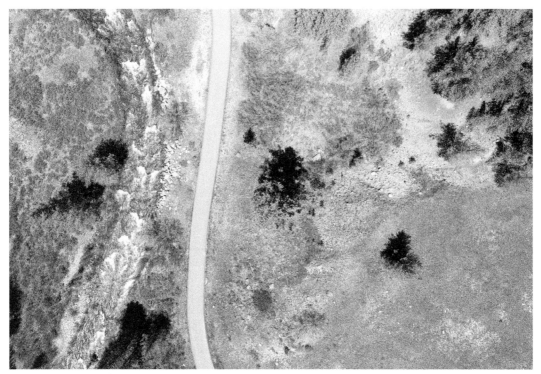

Fig. 4-4: Drones are a great way to gather high-resolution site images early in a project, or even before a project begins. Source: Jon Altschuld/Chinook Landscape Architecture.

photogrammetry means that project design can start earlier in the process. This is a huge advantage for coordination with other disciplines and providing preliminary ideas to clients (Figure 4.4).

Value Added Services

Drone data can be leveraged to create varying types of deliverables, as discussed earlier in the book. Mostly, these are graphics, videos, and plans derived from, or using, drone collected aerial imagery, 3D models, and topographic information. These outputs can be advertised at a greater premium cost; one that clients will pay for. Plus, these are easy products to include in scope and proposal work alongside the basic drone services. In fact, once a firm produces their first graphic derived examples, they should be included with all drone scope work as a separate or additional fee. Give clients the à la carte choice.

It's important to note that these services do require specific skilled labor and software to create. They are covered in greater detail in Part 3, specifically Chapters 8–10.

Here are some examples:

1. *Multimedia videos*. Drone footage and recordings of a site can be used to create high quality videos of existing and proposed conditions and changes for a location. Multimedia videos are a great way to show clients the progress and final designs or can be used for

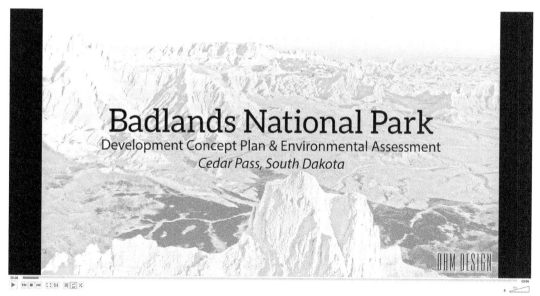

Fig. 4-5: Multimedia videos are a great way to dress up drone footage by adding music, sounds, and captions. Source: DHM Design.

discussions on existing conditions. They can also be used for public and stakeholder outreach, fundraising, social media marketing, and advertising on the firms website and any place else the internet soaks up videos. Most multimedia drone videos can be produced (using Camtasia or Adobe Premiere) within 16–24 billable hours, meaning they are a low labor, high value product (Figure 4.5).

Fig. 4-6: Pix4D 3D point cloud of rockface along US-50 in southwest Colorado. Source: Chinook Landscape Architecture and Yeh and Associates, Inc.

(a)

(b)

Fig. 4-7: (a) and (b) Drone bird's eye photos are easily used as backgrounds for context, merged with 3D models or hand drawings, creating rich, expressive graphics. Source: DHM Design.

2. *3D models*. Photogrammetry can provide 3D point cloud data that can be translated into 3D meshes or what is commonly known as a 3D model. These models can be leveraged in various ways. First, they can be used to produce a concept design of a project location. These can then be merged with drone images and animation. Software like Revit, Rhino, SketchUp, MicroStation, and others work with drone 3D models. It's not difficult to overlay 3D concept models into these drone derived models (Figure 4.7).

3. *Drone video and photo matching*. Merging 3D model images, animations, and even 2D CAD or hand drawn concepts is easy to do with today's software platforms like Adobe Photoshop. These deliverables are more common as drone use spreads in the design world. These images and videos provide excellent ways to convey project design ideas within a site's specific and current context. At DHM, doing work for the National Park Service for Badlands National Park, drone photo matching with 3D models and 2D concept overlays became the primary way for the design team and client to work through the project phases and concepts.

4. *Inspection and analysis*. Analysis of structures, roads, bridges, and slope stabilization have become easier to obtain and view with the use of drones. Using drone photogrammetry derived 3D models and images to inspect walls, roofs, structures, and power lines have brought the cost of these services down, making clients happy while still making a firm money. Firms can reduce liability insurance cost and time spent on many of these services (Figure 4.8).

Fig. 4-8: Zoomed in view of high detail 3D photogrammetric mesh for geohazard mitigation in California. Source: Chinook Landscape Architecture.

Training Costs

Training is divided initially into two parts. The first is training time and costs to obtain the Remote 107 license, required by the FAA to operate a UAV commercially. The FAA has a Remote 107 fact sheet (search remote 107 license). The second part is training time to learn to fly the drone in order to capture the desired data which is covered in Chapter 7.

Beyond the initial costs are training on software for processing data and other data deliverables. These are detailed below as well.

Remote 107 License

The FAA calls this Part 107 but it's commonly referred to as the Remote 107. We strongly recommend subscribing to www.remotepilot101.com to study for the license exam. This website is excellent and has become the go-to site online to study for the drone test. It has excellent content that is regularly updated and also offers renewal study videos. The drone license needs to be renewed every two years.

It takes about 40 hours to study and take the exam. The exam costs $150 at any local official FAA training center and takes about 40 minutes. We recommend setting aside the time to study and take the exam and not spread out the studying over long stretches as that has proven less effective.

Fig. 4-9: Photogrammetry software, while relatively straightforward, does require time for training and understanding in order to use correctly and deliver quality results. Source: Chinook Landscape Architecture and Yeh and Associates, Inc.

Software and Photogrammetry Training Costs

Learning how to use photogrammetry software should be factored into the cost of training. Software like Pix4D or DroneDeploy, while relatively easy, does require some time to learn to use correctly. There are plenty of online videos and tutorials to help assist in the learning process (depending on the software). However, even after the initial learning hours, the ability to generate more accurate photogrammetry requires more time spent learning and playing with the various software; anticipate some additional training costs for each project that involves photogrammetry, even though these can be minimal (Figure 4.9).

Similarly, to work with RTK and GPS survey drones or stations (see Chapter 9) requires greater research and training time. In fact, this starts entering into survey education territory, which can be a deep well of information depending on the firm's goal of using a drone: the accuracy of the information required for a project indicates the continuing research and education costs. Assume similar costs and training for doing 3D concept modeling with higher end modeling and rendering software. One way of realizing these costs is to hire someone proficient in these areas and integrate their experience into the drone flying and data collection goals.

Permissions Costs

In general, there are usually no costs associated with obtaining permission for flying a drone. However, it's worth mentioning in this section to budget for costs in terms of research and time spent in contacting clients, municipalities, or any other important jurisdiction where the drone is going to be flown. For example, to fly the drone in Badlands National Park took an additional 16 hours of work to fill out forms, generate flight maps, and correspond with National Parks officials. For flying over the Garden of the Gods and Pikes Peak took time to formulate flight plans and fill out forms, and also correspond with the City of Colorado Springs. To fly a location in Billings, Montana, near the airport, took 8 hours. Remember to factor in these costs when flying some locations.

Getting Started

This chapter is a road map for implementation. It reviews basic drone features, gear, and flight apps. Included are important skills, understanding hardware, and a review of specific software. These are all qualities required for a well-rounded drone program.

This chapter will review the following:

1. What you need to know about drones (hardware) and how to select a drone.

2. Related computer devices, including tablets, smart devices, laptops, and desktops and how they impact workflow.

3. A detailed review of drone software, what they do, when to use them, and recommendations.

4. A quick review of hardware related to processing drone data like photogrammetry.

AEC Drone Standard Features

There are many drones available on the market. However, drones are not created equal, each one varying in strengths and weaknesses. The authors both use and recommend the DJI Phantom Series and Mavic Pro Series drones by DJI as affordable and capable platforms. At the time of writing, this means the Phantom 4 Pro V2 or the Mavic 2 Pro. To step up from the Phantom/Mavics would mean either the DJI Inspire 2 or the DJI Matrice 200/210 or 600 series (Figure 5.1). All of these are excellent flyers, capture high quality data, are intelligent, and include lots of supporting features and apps, and decent consumer support. Regardless of which drones are available, DJI is still the authors' preferred drone manufacturer for AEC work.

(a)

(b)

(c)

(d)

Fig. 5-1: DJI Phantom 4 Pro (a), Mavic 2 Pro (b), Matrice 210 (c), and Inspire 2 (d), respectively. Source: Jon Altschuld, drone courtesy of DJI Colorado (Centennial, CO).

The drone industry is a large, expanding market, but with surprisingly few manufacturers. DJI, at the time of writing, owns 70% of the global commercial drone market. That looks likely to continue for the foreseeable future due to DJI's range of consumer-professional grade drones, easy to use software, and affordable prices. The Phantom and Mavic series drones are recommended. Specifically the Phantom 4 Pro version 2 and the Mavic Pro version 2 can meet most AEC needs for imagery, videos, and photogrammetry. For the qualities listed below, the Phantom series is used as a reference.

Another alternative is the DJI Phantom RTK drone. The Phantom RTK drone uses the same body as the Phantom 4 drone series. However, this drone is more expensive, because it includes

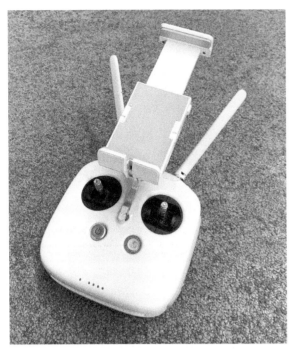

Fig. 5-2: DJI Phantom 4 Pro controller. Source: Jon Altschuld.

RTK GPS receivers (ranging in price from $6,000 to $10,000 depending on the setup). The RTK drone can connect to free public local or national survey systems, for example in the US that would be the CORS network, and can supplement ground control points (required for accurate surveys, see Chapter 9). You can also pay and subscribe to private survey networks.

Flying Skills

Control, stability, and speed are at the heart of a solid AEC drone. Drones should be able to hover accurately in place the instant they take off. The ability to quickly correct and stabilize (because of active GPS connection and positioning) in winds and during high speed flights (stop on a "dime") is important. This is the necessary level of control desired. DJI Mavic, Phantom, Matrice, and Inspire drones include a sport mode to allow for higher speeds to go longer distances while conserving battery life. In addition, drones should be easy to control in any direction of flight and quick to respond to control inputs (Figure 5.2).

The flight controller (what you hold in your hands) is also key to a drone's flying skills. In most cases, you will want to use a flight controller that can plug into your smart phone or tablet. Avoid dedicated controller screens that are built into the controller when possible. Using a plugged in smart device allows for you to run different third party apps (detailed below), which are often not available on flight controllers with dedicated screens. The Phantom and Mavic series allows for both types of controls (smart phone and dedicated screen).

High Quality of Data

Most drones have cameras; however many cheaper ones do not have a gimbal. A gimbal uses an internal gyroscope to keep the camera steady throughout flight, providing a smooth, bump free video and image. For AEC work a drone with a gimbal is essential to generate quality data and professional images/videos. Also essential is the use of a 4K camera, common with most drones. The Phantom and Mavic drones have a fixed camera with a gimbal. They have smooth flight but do not allow for full control of the camera other than pitch (up and down angle). Otherwise the forward direction of the drone dictates the cameras direction. This is fine for AEC projects and does not impede operations or gathering images, video, and data.

The DJI Inspire, the next level up drone from DJI for image/video collection, allows for full independent camera control. The primary pilot operates the drone, while a secondary pilot operates the camera from a separate controller. The Inspire can also be set up to be completely controlled by a single pilot. The Inspire also has several camera and lens options that can be swapped out.

Power and Flight Time

Drone maximum flight times are relatively short. Lithium ion batteries that power drones are slowly improving, but the better drones are constructed from lightweight, durable composite material (for example the DJI Matrice is a carbon fiber: light and near indestructible) with smart power management built in. In the case of the Phantom and Mavic Series, this equates to approximately 27 minutes of flight time in good conditions. At the time of writing, this was the upward limits for flight duration (for a drone of that size) (Figure 5.3). Most automated drone apps will bring the drone home or end a mission when the drone hits 20 or 30% battery life. You always want to have enough power to bring the drone home, and you don't want to be landing at 0% battery. When the battery hits 0% all power is gone. Battery amount is impacted by temperature, wind, and continuous video recording. High winds force the drone to compensate using more battery. Keeping the video recording throughout a flight uses more power. And lower temps cause the battery to dispense energy at a faster rate (ideal battery temperature is 59 °F).

DJI offers smart batteries for their drones. These smart batteries are quick to charge and come with all sorts of data that can be tracked to determine battery health and life. They are stable and help manage power to maximize flight times. There are platforms like Airdata UAV that allow you to read the data from the smart battery, assessing the battery's health, if it has run into any issues and power drains, the number of times it has been charged, and general tracking of the power cells and their health.

Flight Controller Automations

AEC level drones should use smart algorithms to perform almost all functions from flight, power management, and calibration. The drone should be able to complete autonomous flights using apps, return to home when losing connection or another emergency, avoid obstacles, and provide weather conditions warnings. The drone system must be compatible and able to work with apps to perform functions outside of its stock features. These include mounting different

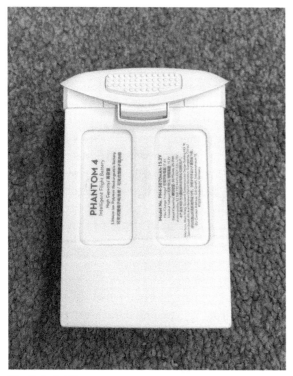

Fig. 5-3: DJI Phantom 4 Pro batteries. Source: Jon Altschuld.

cameras and sensors, and programming complex and repeated flight paths for the drone to follow.

Again, the DJI series of drones allows for all of the above. They include five direction collision sensors and warnings when near objects as well as weather indicator and location warnings, such as if you are near restricted airspace. All of this is valuable in-flight telemetry.

Third Party Apps

To obtain smooth flight paths or gather photogrammetry data and construction information, third party apps are at the heart of drone flying. DJI does make their own flight controller app; this is great for certain types of flights, but third party apps open up a wider range of capabilities. Because DJI has such a large share of the drone market, many apps are developed for DJI drones. These include apps like Litchi, Maps Made Easy Map Pilot, and Pix4Dcapture, all reviewed below. In short, you want apps that support the drone with seamless integration and interface. From the moment you start the app, fly the drone, and bring it back, there should be few to no issues and the drone should do its job of collecting data or recording footage. DJI drones have many app options for filming, capturing photogrammetry data, and maneuvering.

As mentioned, it is often necessary to use a tablet or smart phone to fly a drone to access these apps.

Support and Troubleshooting

Inevitably, you will have to troubleshoot your drone and apps. You should expect that to happen. And expect to go online to look for answers and help, not just from the manufacturer, like DJI, or the app you are using. Companies offer various levels of support. However, with so many people using DJI drones, and associated popular apps, there is a large, organic support network online. Do a search on your issue and review the links. More often than not you will find a solution to the issue.

Drone Packages

There are many types of drones available on the market, and also many drone "packages" (see Chapter 4 for costs) containing the same drone, which can be confusing. However, these packages often offer affordable pricing for a drone, extra batteries, and other accessories.

Below is a list of items to consider including in a starter package. You can purchase drones directly from the DJI website and customize the package. There are also deals through Amazon and other vendors that are more affordable. We recommend a package that includes the items below.

1. *Current DJI Phantom, Mavic, and Inspire drones*: These drones come with one battery and a standard controller.
2. *Two or more extra batteries*: Batteries mean flight time, and the more your operation grows, the more batteries you will need to buy. Having three batteries to start with gives you the ability to make multiple flights at a site and make site visits worthwhile. Six batteries are ideal and should be the goal providing you back up just in case.
3. *Car charger:* to plug in batteries and charge them out in the field. Always allow recently used batteries to fully cool before re-charging.
4. *Multiport charger*: These work with car chargers or regular outlets. It allows you to charge three batteries in sequence. Each battery is charged one at a time but you don't have to babysit changing the batteries as they charge.
5. *Controller lanyard*: This clips into the controller and slips around your neck. This is essential as it ensures you won't drop the controller and it sits comfortably around your neck, taking the weight out of your hands. This is a must have.
6. *Backpack or case*: Another must have item. This allows you to take the drone and accessories just about anywhere and transport them safely.
7. *Waterproof sleeve*: This is for the backpack to keep it dry.
8. *Micro SD card reader*: This allows you to download the videos and images from the drone SD card to a computer. Great to have this out in the field with a laptop and download and review images after flights.
9. *Micro SD card :* 64 GBs of memory is recommended, which can fill up fast. Check the DJI recommended SD card. This is important to ensure the card can correctly sync with the drone to effectively record videos and photos. With HD ultra high writing speed (look for UHS Class 3 cards) these cards can keep up with the camera taking snapshots or recording videos.

There are other drone companies, like Yuneec and Parrot, as well as fixed wing drones. However, for the purposes of this book, as already stated, we use DJI Phantom drones as they are well tested, reliable, and affordable. Fixed wing drones have seen great success in certain applications, particularly for very large areas that do not need ultra high – resolution data. However, they are much more expensive than quadcopter drones and so their use has been more limited to projects that can support their cost and use.

Applications and Software

There are many apps that can be discussed for drone use. Chapter 9 goes into greater detail (and examples) of some of the recommended software to use with the DJI series of drones; much of the software listed can be used with a wide range of drones. We are going to focus on some popular ones that have regular updates and support. The apps are subdivided into multiple categories and subcategories based on use and function.

Below are the three main categories of apps/software and their associated subcategories, with a list of useful apps worth trying.

Device Apps

These are apps for smart phone and tablets that are available from the Apple or android store. Many are free, have a free trial, or are low cost. There is a broad range of apps; the ones listed below are recommended for most AEC projects. DJI and many other drone manufacturers have their own flying app as well. For example, the DJI GO app, allows for the interface and control of a drone through a smart device.

When it comes to these types of apps, we strongly recommend experimentation and practice. There are always news apps on the market, and given the relatively young age of the drone industry, there is constant development and improvement. Similarly, don't stick to one app for any discreet task. Load up the various competing apps, check conditions on one or all of them and be flexible about which ones you use. Online forums and communities are also a great resource to see how others are using the available apps.

Categories

There are five discreet categories apps provide functions for. Below are detailed descriptions for each category and recommended apps readers can start out with and use in their practice. The apps below can be found in the Apple and Android stores.

1. Flight conditions
2. Video and photo recording
3. Photogrammetry
4. Insurance
5. Online portals

Flight Conditions

Flight conditions relate to weather, FAA airspace limits, and location specific rules and restrictions. For example, some apps will indicate if you're near a heliport, restricted airspace or no fly zone, wind speed, visibility, etc. Many apps overlap functions providing weather and location information. It is important to note that sometimes these apps are not always accurate, for weather conditions in particular. It is most important to assess a flying location first hand and determine if it is ok to fly. It is the responsibility of the pilot to use the standard approach outlined by the FAA (part of the Remote 107 exam and rules, see Chapter 6) to determine flight conditions. These apps are not intended as a replacement.

- ▶ *AirMap*: To check local flying rules, proximities, and restrictions, AirMap is a current app of choice. It provides weather conditions and the ability to create flight plans and logs. It's an excellent example of a multifaceted flight conditions map. AirMap also integrates UAV laws for over 20 countries. Furthermore, AirMap provides access to the FAA LAANC permissions system to obtain permissions to fly the drone in US restricted airspace.

- ▶ *SkyVector*: This is actually a website, so it can be used on either a desktop computer or a mobile device. SkyVector is an interactive web map that allows users to view aeronautical sectional charts. These depict the FAA airspaces that you are required to adhere to when flying commercially (Figure 5.4).

- ▶ *UAV Forecast*: Excellent holistic app. A simple message at the top of the app indicates if you're good or not good to fly.

- ▶ *Windy*: Detailed, up the minute wind information, which is crucial for flying in locations with higher or unpredictable wind conditions. Windy provides info from local weather stations and also wind and gust forecasts from multiple weather forecasting models (Figure 5.5).

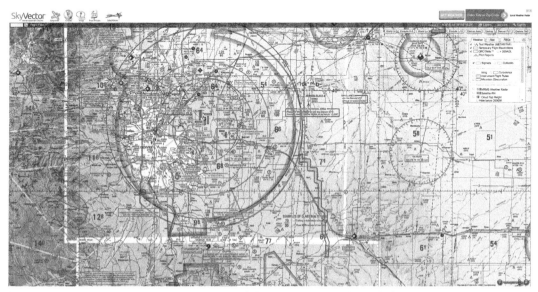

Fig. 5-4: FAA Sectional Charts and Notices to Airmen (NOTAMs) are viewable for free in SkyVector. Source: SkyVector.com.

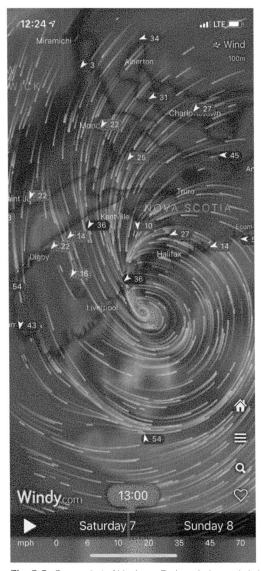

Fig. 5-5: Screenshot of Hurricane Dorian wind speeds in Windy iPhone app.

Video and Photo Recording

Recording drone footage, photos, or videos is a very standard affair. There are many apps to select from, and, like flight condition apps, many apps overlap in function allowing for various recording of data. Listed below are apps that assist with recording straight photography and video footage and not data or photogrammetry data collection (see next category).

▶ *DJI GO:* The DJI basic flight app is a quick and easy way to record photos and videos. Simply click on the button and it will start recording. Tap the screen during a video

Fig. 5-6: DJI Go app screenshot. Source: Jon Altschuld.

recording flight and you can capture a photo as well. To add more control over photo and video recordings, DJI Go offers many custom, manual features including changing the photo or recording ratio (4:3 versus 16:9 for example), color levels, capturing RAW photos or full channel videos (DLOG and LUT). There is a plethora of options to select from. This is a solid app to use and if you use a DJI drone, you will learn to use these features (Figure 5.6).

▶ *Litchi:* This is the recommended app for recording video and is becoming an industry standard. Simple to use with excellent features if you want to capture smooth, consistent, professional grade footage without being a pro; this is a must use app. Litchi has a very useful online portal best accessed through a laptop or desktop. It allows for precise creation of flight paths for the drone, including the adjustment of parameters such as flight speed, camera direction, focus, smooth gimbal movements, being able to take snapshot photos while recording video and more. Once the flight paths are created online, the device app can be easily accessed (just log into your app) and with a click of the button, the drone will take off, fly, record, and land on its own. These flight paths can be created directly in the app as well, either before flight or during flight.

For many people learning to fly a drone, capturing solid video footage is a challenge. Jerky and quick flight motions make videos hard to watch. Litchi solves that issue and then some. Litchi is reviewed in greater detail in Chapter 8.

Photogrammetry

There are several options for the capture of photogrammetry quality images and processing (photogrammetry is covered in detail in Chapter 9). Because photogrammetry uses photos, any app that captures photos can be used to collect the necessary data for photogrammetry. However, photogrammetry requires a large number of photos with regular overlaps, and there are a number of apps that automate the capture of these photos. Each app below has an associated web portal or software to process the data, but they are interchangeable. For example, photographs collected with the Maps Made Easy Map Pilot app can be processed with Pix4D or any other photogrammetry software.

These apps are similar in use. Users define the area to fly through the device app. These are usually flight grids or circular flight paths. The app then uploads the flight plan to the drone and automates the entire flight process. The apps allow users to set the scanning resolution (from 1/2 inch to 1 inch, etc.), provide estimated flight times, adjust the scanning paths, etc.

Once the flight is completed, users are required to upload the collected images either to the apps cloud service or to a local software. This is detailed more below for each app.

The authors recommend all the apps listed below for this type of work. Each has its own way of working and it will come down to preference and comfort.

▶ *Maps Made Easy Map Pilot*: In the Maps Made Easy Map Pilot app, simply touch the screen to create a boundary area for your location and the app does the rest. You can set the resolution based on flight height – the lower the drone flight elevation, the longer the flight and the more detailed the resolution. The app provides full flight automation.

Once the flight data and images are recorded, either upload the images through the Maps Made Easy website where the images are processed, or insert the photos into other photogrammetry software. Maps Made Easy processing fees are based on the total number of images and output that are required. The website and app also allow for pre-flight planning and saved paths to determine flight duration and locations.

The website is easy to use and includes quick tutorials (Figure 5.7).

▶ *Pix4Dcapture:* The Pix4Dcapture app provides multiple flight options and data recording: 2D flight grids for quick, high-resolution aerials, oval flight paths, and detailed grids for 3D generated data. The flight paths, once inputted into Pix4Dcapture, are fully automated. Flight paths can be set up during pre-flight stages and run once in the field (Figure 5.8).

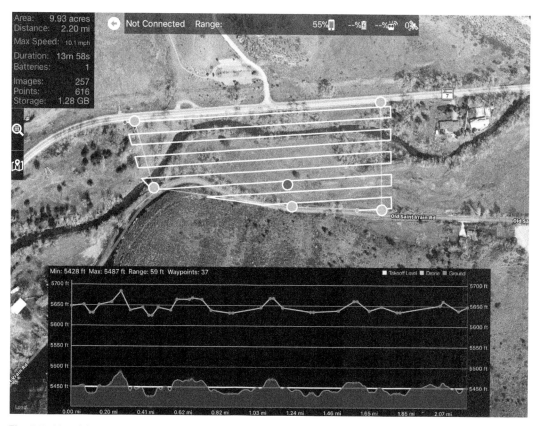

Fig. 5-7: Maps Made Easy app screenshot. Source: Jon Altschuld.

The data can be uploaded to Pix4D's cloud or can be run locally through the Pix4D Mapper software. The Pix4D online and desktop software require more nuance and investment in time to learn to use but provide some of the best outputs and more control than fully automated processing such as Maps Made Easy (such as adding Ground Control for geo-rectified data). Classified point cloud and 3D model data, AutoCAD/DWG contour lines, and elevations maps are all possible outputs. Running the software to process the data locally is faster than using the provided cloud service but requires a powerful computer (see hardware below).

▶ *DroneDeploy:* This is a popular competitor to Pix4D and offers the same types of features including automated flight paths, as well as online and local software to process the data. For most users, choosing between the two will come down to personal preference (Figure 5.9).

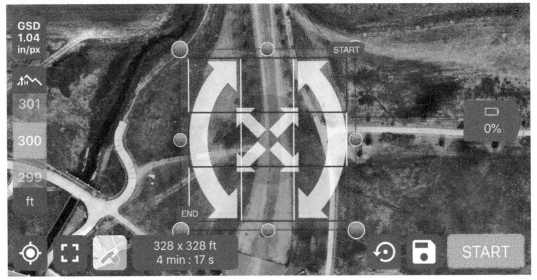

Fig. 5-8: Pix4Dcapture app screenshot. Source: Daniel Tal.

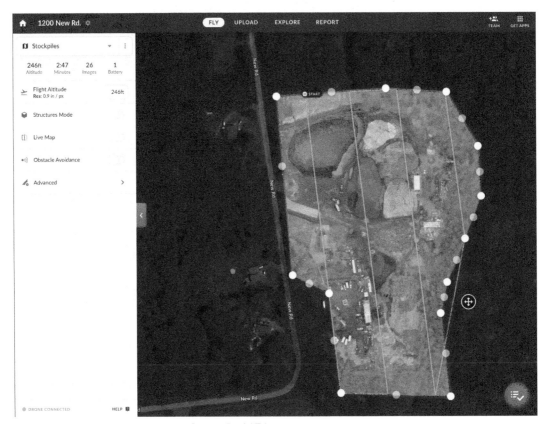

Fig. 5-9: DroneDeploy app screenshot. Source: Daniel Tal.

Insurance

Liability coverage is reviewed in greater detail in Chapter 6. Two apps, Verifly and Sky-watch, provide spot coverage for flying a drone site. Charging as little as $5 an hour, app users can obtain coverage for a specific flight location. Coverage prices vary, as do the amount covered (for example up to $1 million versus $500,000). These apps are great for people who do not carry comprehensive business insurance. Both these apps also offer hull insurance in addition to liability coverage.

In addition, the apps indicate the conditions in the flying location and how they relate to the cost of coverage. A dense urban area near an airport might cost between $25 and $30 an hour and have a lower policy liability.

Otherwise, there is not much difference between the two apps so this comes down to user preference. For those starting out (including hobbyists) we recommend using these apps until you need to obtain comprehensive insurance.

Online Portals

Some phone and tablet apps have associated online portals that work in tandem or have synergistic uses. For example, the Litchi app has an online portal that can be used from a laptop or desktop. This portal can be used to pre-plan flight paths, camera angles, and drone speed, among other features. These missions can be saved and then uploaded to the Litchi device app. Pix4D, DroneDeploy, and Maps Made Easy have a flying app for your device and powerful online and desktop software where data is uploaded and processed.

Desktop Software

Desktop and laptop software are usually powerful platforms that provide a variety of needs ranging from video and photo editing to 3D modeling and photogrammetry processing to be used on a laptop or desktop to process images, videos, and data. Many of these platforms require more robust computer hardware or specific requirements to run efficiently. This is reviewed below under hardware.

Photogrammetry

To process photogrammetry images (see Chapter 9) requires accessing software through a web page and cloud service or dedicated local software. This is noted above for Maps Made Easy, Pix4D and DroneDeploy. Chapter 9 discusses the details of processing photogrammetric data, as well as several software options.

Photo Matching and Editing

Adobe Photoshop is an industry standard software that is used to touch up drone photos. A photo editing software is required to produce drone photo matching which is the combination of 3D models, 2D plan graphics, and 2D CAD plans over drone images and aerials.

Multimedia and Video Editing

Listed below is the recommended video editing software readily available at most firms.

1. Camtasia, by TechSmith, is an easy to use, affordable, stable video editing software. It allows for captions, titles, graphics, animations, and many video outputs (like MP4).

2. Adobe Premiere Pro, like Photoshop, comes as part of the Adobe Creative Cloud package. This is relatively easy to use, full featured editing software. It can handle large video files and provides many options for touching up videos.

3. Adobe After Effects: Also part of the Adobe Creative Cloud, this high-powered video editing software has been used for years to post-produce Hollywood films and special effects. It is a required to be able to merge drone footage with animated 3D models. The process is not simple and requires a knowledged user to perform.

Computer Hardware

When talking about computer hardware, we are typically referring to laptops or desktops to crunch drone data, like images, videos, and photogrammetry. For laptops and desktops, the three most important elements are the CPU, for example Intel-i9 or better chips, RAM, and graphics card or GPU. There are three basic levels of computers you will want to process and work with drone data:

Images and simple videos: If you only want the images and basic videos, almost any device these days can handle that. Even basic to more advanced photo editing does not require anything beyond a simple laptop.

Multimedia and 3D visualization: Merging drone footage with 3D models is an effective presentation graphic. Similarly, processing high-resolution 4K drone video requires solid processing, RAM, and graphic cards. These computers usually require top end, multi-core processors, a minimum of 24–32 gigs of RAM and the best (recommended) NVidia GeForce graphics card (on the market). These computers can be both laptops or desktops.

Many gaming computers fit this bill but it is still best to build a custom computer, through a local computer service shop or go online and customize a system through gaming/industry specific computer websites like iBuyPower, XI computers, Falcon, Dell, and others. The average cost for these computers and laptops is around $1,000–4,000.

Photogrammetry: In many cases the multimedia and 3D computers are enough to do photogrammetry, but very detailed, large scale data processing times can be long. Most photogrammetry software uses ALL of the computer's resources in various phases to produce a point cloud, 3D mesh, and classified data (reviewed in Chapter 3). Some companies build photogrammetry processing computers. In simplest terms these computers max out RAM (for example 128 gigs), use a top rated or top tier CPU chip and motherboard, and the highest end graphics card (if not multiple cards). These computers are ideal if you process at least 1 to 2 large drone datasets per week. The typical starting cost is no lower then $3,000. Puget Systems (https://www.pugetsystems.com/solutions/engineering/photogrammetry. php) builds photogrammetry specific computers. The other option is to build a computer from off the shelf parts (that you can get from Micro Center for example).

CPU

The CPU, or processor, plays the lead role in processing the millions of data points and pixels from all of the photographs. CPUs are defined by their speed (measured in GHz), and the number of processing cores they have. Most, if not all, desktop photogrammetry software can utilize multi-core processors. As a result, it is best to find a balance between high speed and high core count. Many processors not only have multiple cores, but also multiple threads. This means that each core can handle multiple processes at a time. For example, you may see processors listed as 4 core/8 thread. This means they have four cores, but each core can act as a dual core processor.

GPU

The GPU, or video card, is traditionally used to process what is being shown on monitors or displays. However, they have been used more recently as a powerful way to process large amounts of data, similar to the CPU. Different software can utilize the GPU in different ways, so it is important to check what each software company recommends. An important factor to any GPU is the amount of dedicated memory on the card. This memory is dedicated only to the GPU processes, as opposed to being shared with the rest of the computer components. The more dedicated memory, the better. Of course, the more dedicated memory, the costlier the card will be.

RAM

The RAM memory is the memory that the CPU uses to complete all of the processes. Although sometimes confused with storage memory (hard drives), this memory is completely separate and different. RAM memory is used during processes and is freed up when processes are not occurring. RAM can be upgraded (both by speed and amount), but the maximum amount of memory is typically limited by the computer's motherboard capabilities.

Storage Memory

The final limiting component is storage memory. This is where the actual files are stored and is typically either a hard drive or a server. While this is less limiting than the CPU, GPU, and RAM, it is still an important consideration for two reasons. First, certain photogrammetry steps will run quicker if the storage drive it is working off is a fast drive. There are multiple options ranging from mechanical hard drives (slow), to SSD solid state drives (faster), to NVME M.2 (very fast) cards. The second reason is that photogrammetry files and their associated photographs take up a lot of storage space. Efficiently organizing the files and having a set process/procedure for archiving older projects is key to running a successful drone photogrammetry process.

It is not recommended to use laptops with photogrammetry. The process makes a computer run hot and laptops, even with additional fans for ventilation, can be overheated by this process. Desktops, custom ones in particular with larger towers, more fans and better ventilation, are better suited to transfer and reduce heat.

Documentation, Permissions, and License

This chapter will review the licensure and documentation important to operating a drone for commercial use. Most of the requirements are pretty straightforward and in some instances are just recommended. It is worth implementing even the suggested documents and record keeping to maintain effective, safe, and legal drone business.

Professional Etiquette for Professional Practice

Although professional use of drones is arguably in its infancy, it is progressing quickly and drones are being used in a very wide range of fields. As a result, drone use is growing more rapidly than standards of practice, and legal regulations in particular, are being developed. This chapter will detail the legal requirements, the gray areas, and the common sense practices for using drones in professional settings. Always err on the side of caution and safety when unsure of how a regulation applies to your situation, or if there is not yet a regulation for your situation. Because drones are visible and exciting, one user acting irresponsibly has a large effect on how the overall industry is viewed. The more drone users follow a professional standard of conduct the better the perception of the industry in the eyes of the public.

The Legal Rules

For the purposes of this book, we will focus on legal rules and regulations within the United States of America. Within the US, the Federal Aviation Administration (FAA) is the regulating and enforcing agency for aerospace and airspace. Although each country has it's own regulations, many are very similar to the FAA's, and so it is worthwhile to be familiar with the FAA regulations. Their mission is "to provide the safest, most efficient aerospace system in the

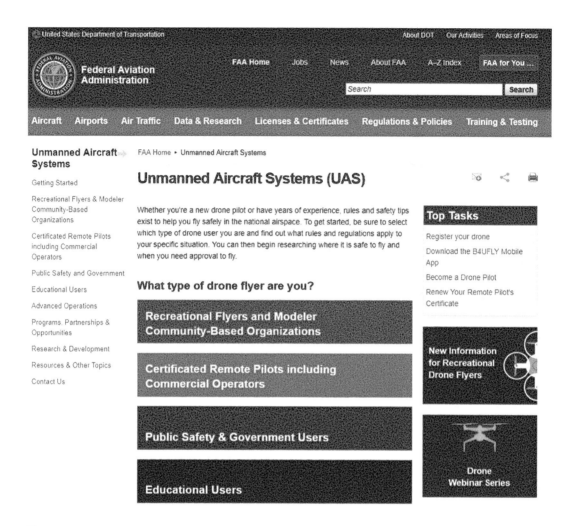

Fig. 6-1: Screenshot of the FAA website for Unmanned Aircraft Systems. Source: FAA.gov.

world" (https://www.faa.gov/about/mission/) (Figure 6.1). To understand how the FAA regulates drone use, we have to first take a brief look at the history of drone regulations.

Historically, the FAA has been much more focused on manned aviation operations than unmanned, but they do have jurisdiction over both. Prior to the widespread adoption of drone use, the FAA's unmanned regulations pertained mostly to hobbyists and commercial operators with very expensive fixed wing aircraft. The number of operators (hobby and commercial) was relatively small and the potential for conflict between manned and unmanned operations was also relatively small.

As drones became cheaper to buy and easier to fly, the FAA has had to create new regulations for their use, particularly commercial use. These regulations started as very cumbersome and difficult to obtain Certificate of Authorizations (COAs) in the later 2000s, in which a public entity had to be a sponsor of each individual application. The FAA Modernization and Reform Act of 2012 slightly improved this process by instead requiring commercial users to apply for a 333 exemption. These exemptions typically took six months or so to obtain, but they were less restrictive and more commercial entities were able to secure them. However, most 333 exemptions still required the operator to obtain an FAA Airman's Certificate. There are various types of Airman's Certificate, such as private pilot's licenses. One of the more common certificates that drone pilots were obtaining during the "333 exemption era" was a lighter-than-air (LTA), or hot air balloon, pilot license. As you can imagine, this was a big limitation to the commercial drone industry.

While the 333 exemptions seemed over-limiting at the time, they provided a temporary set of regulations while the FAA developed a permanent set of rules specifically for commercial drone use. In 2016, the Small Unmanned Aircraft Rule, or Part 107 for short, was released. Part 107 details rules and regulations for commercial drone use, and it also introduced a new category of airman's certificate: the Remote Pilot Certificate.

Part 107

In order to fly a drone commercially, one must adhere to the Small Unmanned Aircraft Rule, also known as 14 CFR Part 107, or just Part 107. To fly under Part 107, one must test for and receive a Remote Pilot Certificate, register the UAS/drone, and follow all Part 107 rules (Figure 6.2). Under FAA rules, a *small* unmanned aircraft weighs less than 55 pounds, including payload.

Fig. 6-2: Example of an FAA Remote Pilot Certificate card. Source: Jon Altschuld.

The first step to flying commercially is to study for, and pass, the aeronautical knowledge test for a Remote Pilot Certificate. Eligibility requirements are fairly simple and include being at least 16 years old, being able to read, speak, write, and understand English, and being in a physical and mental condition to safely operate a drone. If you meet these prerequisites, you can register for the knowledge test. The test is an electronic, multiple choice exam, and you receive results upon completion. Study material for the exam is available from a wide variety of sources online, including the FAA's "Safety Team" (FAASTeam) website. Another great source of study material is RemotePilot101.com, which provides study courses and sample test questions for a fee.

Following completion of the exam, one must create an IACRA account and wait for the FAA systems to receive the test results (Figure 6.3). Once this is done, a plastic certification card is sent to the recipient through the mail. The Remote Pilot Certificate is good for two years, at which point a recurrent test must be taken to maintain the certificate. Both the initial and recurrent test must be scheduled and taken at an FAA-approved knowledge testing center (KTC).

Although the Remote Pilot Certificate, or license, is one of the most talked about steps to flying commercially, it is not the only one. The aircraft must be registered with the FAA, and the pilot must still adhere to all Part 107 rules. Registration is a simple and straightforward process, and can usually be completed in under 10 minutes on the FAA website.

A full listing of the Part 107 rules can be found online, but they include:

▶ Aircraft must weigh less than 55 pounds, including payload

▶ Fly in Class G airspace

▶ Aircraft must remain within visual line of sight (VLOS) of pilot

▶ Do not fly over people

▶ Do not fly over moving traffic

▶ Do not fly above 400 feet AGL

For many of the Part 107 rules, a waiver or authorization can be applied for. For example, although the rule is to only fly within Class G airspace, you can apply for an authorization to fly within any airspace, for example Class B, C, or E which are restricted locations. The FAA is under no obligation to grant such an authorization, but with proper planning and safety provisions, they are regularly granted. Depending on the complexity of the rule, the requested action, and the required review, turnaround on waivers and authorizations vary from minutes to several months.

Permissions – LAANC

Airspace authorizations in particular have been prioritized by the FAA, and they have created the Low Altitude Authorization and Notification Capability, or LAANC, system. LAANC is an automated authorization system in which pilots can apply for immediate authorizations within many airspaces around airports. The LAANC does not give blanket access to all airspaces, but it is a huge improvement over the previous case-by-case review method. For many airports, there is still a no fly zone immediately around it. Further out, there will typically be areas that can

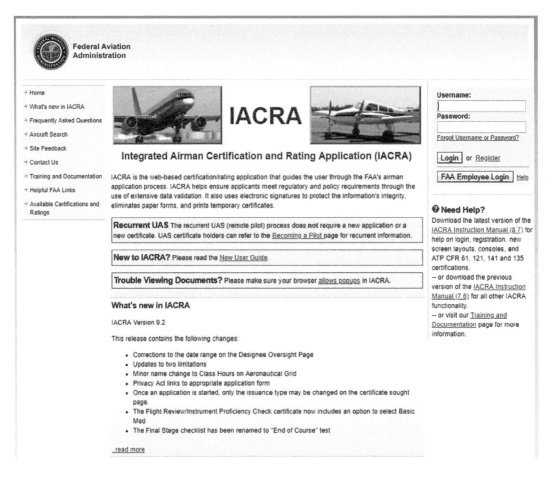

Fig. 6-3: Screenshot of IACRA website. Source: iacra.faa.gov.

be flown up to 100 feet AGL with an automated authorization, followed by areas further out in which flights up to 400 feet AGL can be flown with automated authorization.

The LAANC system can be accessed using apps like AirMap and Skyward (they have a website as well, Figure 6.4) or through the LAANC website. You will be required to register and provide information about your flight, location, and time. Authorization, if approved, can be quick. The process is seamless. Please note, this is not the same as applying for an FAA wavier. The FAA waivers are for deviating from Part 107 flight rules.

Fig. 6-4: (a) Screenshot of the United States in the Skyward.io LAANC interactive map interface. Yellow areas have certain airspace restrictions that you can have authorizations automatically approved for. Red areas are either not part of the LAANC system or are more restrictive. These will require the traditional FAA waiver process. (b) Screenshot of the LAANC interactive map in Skyward.io. By clicking on each yellow area, the system shows what ceiling altitude an automated approval can be applied for. In this case, the center area containing the airport itself will not allow for any automated authorizations (ceiling of 0 feet). The next "layer" out will allow for automated authorizations up to a 100 feet ceiling (from ground, not from structures), and the furthest "layer" will allow for automated authorizations up to a 400 feet ceiling. Source: Skyward.io.

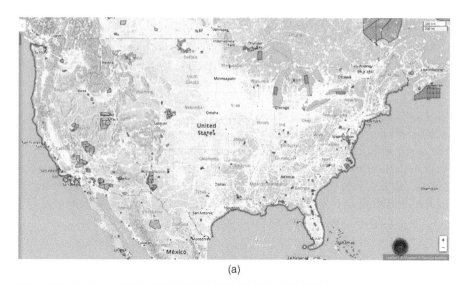

(a)

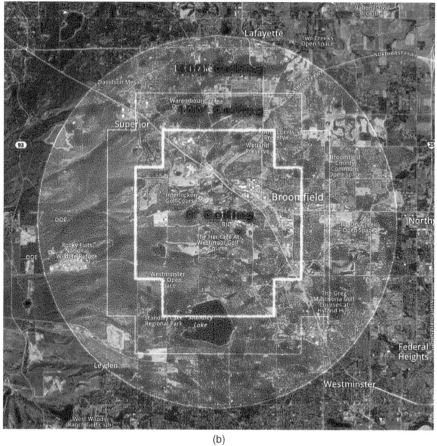

(b)

Permissions – State, Agency, County, and City Regulations

In addition to the FAA rules, many states, agencies, counties, and cities have tried to implement their own drone use rules. Generally, these organizations do not actually have jurisdiction over the airspace above their property, and the FAA is the governing body. However, they do have jurisdiction over what occurs *on* their property and many have enacted policies regulating the takeoff and landing of drones on their property in an effort to regulate drone use. While this has created a patchwork of rules that can be confusing to keep up with, it is always prudent to research and sometimes even contact the jurisdiction in which you plan to fly. Adhering to local regulations may require extra paperwork or applications, but adhering to their rules will help to establish yourself as a professional, rather than a random drone operator trying to skirt the rules. Often in the AEC industries, the local municipality or agency is also the client, and it's never a good idea to ignore the clients' regulations (Figure 6.5),

Permissions – DJI FlySafe and Geo Zones

DJI has its own website to restrict drones from flying in specific locations, in particular within five miles of an airport or similar regulations. It is important to check any location you will fly with a DJI drone at https://www.dji.com/flysafe/geo-map or do a search for DJI geo-map. This process can be fairly involved so it's crucial to do this before going to any mission site in order to ensure you can fly.

You can find the exact location you plan on flying and if there is a restriction to either take off or to a vertical flight limitation. If there are any restrictions on taking off or if you are limited to a specific flight height that is below the mission requirements, you will need to unlock the drone for that mission through the DJI Geo Zone Map website. Failure to do so can cause the drone to refuse take off and fly as these locations are coded into the drone's firmware (Figure 6.6). It is important to note that DJI REQUIRES drone pilots to get approval from the FAA (if within an FAA restricted area – they do not always align with DJI Geo Zones) and submit that paperwork with the DJI request to unlock the drone. In many instances getting both LAANC and DJI permissions to unlock an area can be done in under an hour.

Insurance

To fly a drone commercially, the FAA does *not* require liability insurance. Even though it's not required, it's tough to imagine why a professional drone user would not purchase liability insurance. Think about all of the other aspects of your business, and how each and every one of them is probably covered by either commercial general liability, errors and emissions, workmen's compensation, unemployment, commercial automotive, or an umbrella policy. Out of all those aspects, having an accident while operating a drone is probably more likely than many

Fig. 6-5: (a) follow flight plans and (b) follow flight paths. Some municipalities and agencies will require in depth applications and information in order to take off or land from their property. For this project, the application required detailed flight plans including flight paths, operator areas, and flight elevations/altitudes, as well as photographs of the aircraft and copies of insurance, pre-flight checklists, and remote pilot certificates. Source: Jon Altschuld.

Remote Pilot is FAA licensed.

Both the operating company and the general contractor hold liability insurance specifically for UAS/Drone operations.

UAS will be within Visual Line-of-Sight of the Remote Pilot at all times during flight.

UAS operating company maintains flight and maintenance logs.

UAS operating company performs pre-flight checklist before every flight.

Although some flight paths may show brief periods of flight over minor roadways, flights will not be conducted over moving traffic. If roadway has light/minor traffic, the UAS operation will occur during times of no traffic. If traffic is consistent and/or too heavy to perform the operation safely, the flight plan will be adjusted to avoid flying over the roadway.

Exact flight paths may be field adjusted to account for structures, vegetation, traffic, events, etc.

All flights will be flown at approximately 200' AGL (above ground level).

Flight #/path; detailed flight path and elevation on following pages.

(a)

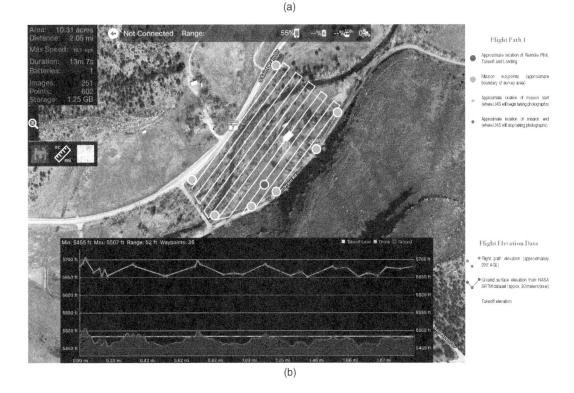

Area: 10.31 acres
Distance: 2.05 mi
Max Speed: 16.1 mph
Duration: 13m 7s
Batteries: 1
Images: 251
Points: 602
Storage: 1.25 GB

Flight Path 1

Approximate location of Remote Pilot, Takeoff and Landing

Mission waypoints (approximate boundary of survey area)

Approximate location of mission start (where UAS will begin taking photographs)

Approximate location of mission end (where UAS will stop taking photographs)

Min: 5455 ft Max: 5507 ft Range: 52 ft Waypoints: 36   ■ Takeoff Level ■ Drone □ Ground

Flight Elevation Data

Flight path elevation (approximately 200' AGL)

Ground surface elevation from NASA SRTM dataset (appox. 30 meters/pixel)

Takeoff elevation

(b)

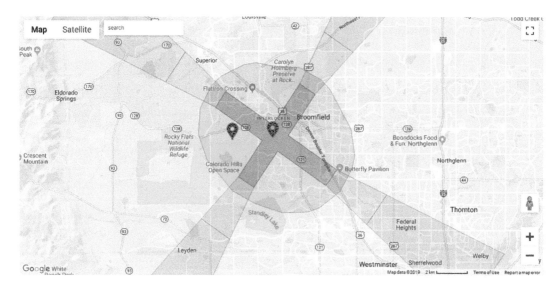

Fig. 6-6: Screenshot of DJI's Geo Zone Map; this is the same airport shown in Figure 6.4b. It is important to note that DJI's Geo Zones do not align with the FAA/LAANC boundaries, and in many cases, are more restrictive. Even though you may not need an authorization from the FAA to fly within the outer "gray" areas shown in the DJI map, the DJI app will limit your drone's maximum height within this area (if flying a DJI drone of course). Source: dji.com.

things that those would cover. Drones can lose radio signal, batteries can fail, propellers can break, or pilots can simply make an error and cause an accident.

Liability insurance for drone operations is generally very affordable when compared to other liability insurance premiums. Many insurance providers are now offering drone liability insurance packaged with their other policies, whereas companies used to have to purchase a special policy just for the drone operations, often with a separate provider. Most drone insurance policies have options to add hull insurance to cover the cost of repairing or replacing the drone itself, in addition to any damage the drone causes to property or people.

Documentation

Once you are legally able to fly a drone commercially, it is important to consider what documentation you should keep with you when operating the drone. Having your remote pilot certificate and UAS registration with you is a must, as you are required to present these items if requested by a governing agent. If you carry insurance, it's also a good idea to keep a copy of your policy and contact info for your agent in the case of an accident. If you are flying under an FAA Authorization or Waiver, you need to have a copy of those with you as well. These documents are also great to have on hand (including a business card) when approached by someone from the public who is curious as to why are you flying a drone.

It is recommended (though not required) to keep a log of your flight operations. There are many ways to do this, ranging from manually writing down records of flight times, battery percentages, weather conditions, etc., to using logging management systems to keep track of flights and conditions. In addition to logging your operations, it can be a good idea to log or track your maintenance. It is required that an operator keeps the aircraft properly maintained according to

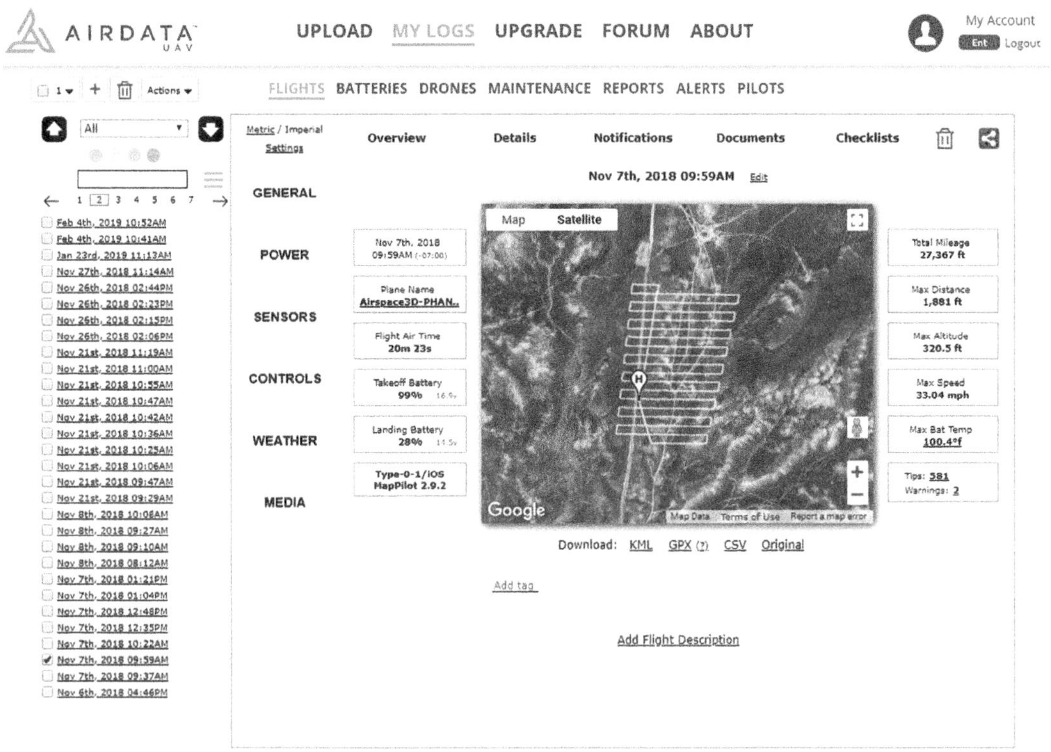

Fig. 6-7: Screenshot of the AirData UAV Dashboard and the general overview of a specific flight. Source: airdata.com.

manufacturer's recommendations. One item regarding maintenance that is often overlooked is software updates. Flying a drone with an out of date control device (tablet) operating system or flight app (DJI Go4, Litchi, etc.) actually qualifies as improper maintenance.

Although there are several options, one of the best flight logging and maintenance tracking systems available is AirData UAV. AirData UAV is an online management system that can automatically sync with many flight apps, including DJI Go, Litchi, DroneDeploy, MapPilot (Maps Made Easy), Pix4D, and others. Through this syncing, all flights can be easily uploaded to your online account where you can view a lot of important flight data in your internet browser. AirData UAV pulls from all of the data the drone collects, providing you with a much more robust perspective of your flights and equipment (Figure 6.7).

For example, AirData UAV reads the battery information from every flight, but not just the basic "starting percentage" and "ending percentage" values that are visible in more standard flight viewers. AirData UAV also displays the voltage deviations for each battery cell, which is a great indicator of battery health. Each DJI Phantom 4 battery contains four cells, and while some deviations are normal, if a cell has more than 10 major deviations (deviating more than 0.07v from the other cells), it is a good indicator that the battery will fail soon. AirData UAV also identifies the battery used in each flight, and tracks trends in battery health such as battery life and capacity. These generally decline over the life of a battery and knowing when to retire a battery is an important part of maintenance (Figure 6.8).

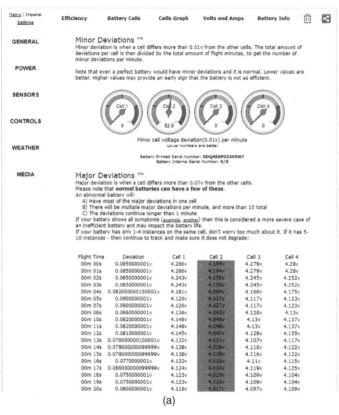

(a)

Battery Cell Deviations over time: each bar represents a flight, ordered by date.
Click on each bar for more info. Here is an example of a problematic battery.

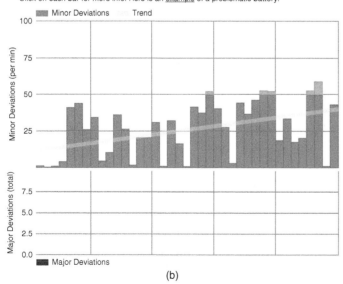

(b)

AirData UAV also includes an HD flight player, which allows you to replay each flight over a map, along with detailed telemetry and alerts – similar to what you would see during the flight, but with additional information. AirData UAV has developed a 3D aerodynamic model for each aircraft type, enabling in-flight wind calculations, which can be viewed on a map, or as a wind–altitude profile. These are helpful when flying in close proximity to structures, as it allows for planning the safest approach, or in repeating flights, as a way to set the proper cruising altitude to maximize flight time (Figure 6.9).

Another valuable feature of AirData UAV is the ability to schedule regular maintenance for aircraft and batteries. Depending on the tier of membership, you can either adhere to recommended maintenance schedules and routines, or you can create custom maintenance schedules. Users can also create indicator thresholds to trigger alerts, and generate customized reports for internal use or regulatory requirements.

Overall, the legal requirements to fly a drone for commercial use are not a barrier to implementing a drone practice into a business. The perception of the legal requirements being a barrier to entry is often greater than the actual requirements themselves. That being said, the extremely quick growth of the consumer and prosumer drone markets, combined with this perception, has resulted in many "professional" drone users skirting the rules and federal policies. In order for drones to be more widely accepted into professional practice, commercial drone users *must* adhere to laws and regulations, similar to any other professional practice.

Fig. 6-9: Screenshot of the AirData UAV HD flight player, in which you can view every control input and notification as well as detailed information and statistics during that notification. Source: airdata.com.

Fig. 6.8: (a) Screenshot of AirData UAV data on a drone battery. This was a brand new battery and without analyzing the AirData UAV information there would have been no signs that this battery was defective and likely to fail. (b) AirData UAV also tracks trends across time. Because most drone batteries are "smart batteries," AirData UAV can automatically identify each specific battery. This chart shows that this particular battery is developing a trend of having more cell deviations as its life progresses. Source: airdata.com.

Best Practices
for Flying Drones

So far, Part 2 of this book has explained how to develop a plan for a drone program, "sell" it to your company, and choose your equipment and software. You have also learned about the legal requirements for flying a drone commercially. The last remaining piece of the puzzle is learning to fly the drone.

Although drones are notoriously easy to fly (this is, in fact, one of the reasons they have gained such widespread use so quickly), there is a big difference between flying a drone, and flying a drone for efficient business practice. For a business, flying a drone requires an on-going interest in bettering your flying skills, establishing company-wide standards of practice, and standards of recording (Figure 7.1).

The Flying Mindset

Learning to fly a drone responsibly involves more than simply starting up the drone and taking off. It requires the mindset and awareness of all the different pre-flight checks and planning. This is similar to a standard airplane pilot's pre-flight checklist and training.

For example, prior to any flight, it is important to identify the purpose and parameters of the flight. While this may seem obvious for project specific flights, it also holds true for training flights, even your very first flight. Prior to taking off, time should be spent researching the drone's controls, how the flight controller app works, and what to expect during the flight.

Some basic considerations that are part of learning to fly a drone include:

1. How do you start the motors?

2. How do you stop the motors?

Fig. 7-1: Drone operations can be fun, but integrating them into professional practice requires a commitment to training, maintenance, and pre-flight planning. Source: Jon Altschuld.

3. What happens when certain controller buttons are pressed?

4. How do you maneuver the drone up, down, sideways, point it in specific directions, etc.?

5. What should you do if something unexpected happens?

6. Where can you land the drone if something unexpected happens?

7. What apps do I use to fly the drone?

8. What is the basic equipment (drone, controller, screen, batteries, cords) needed to fly?

9. How do I carry all the equipment?

Much of this information can be found in the manufacturer's documentation that comes with the drone, or you can search online resources like YouTube and user forums.

Drone Flying Apps

Over time, the more you fly, the more you'll likely use multiple flight controller apps, each with its own unique interface and capabilities. Whether using the DJI GO 4 app (Figure 7.2), Litchi, Map Pilot, or any other flight controller app, it is important to familiarize yourself with how each one works and how you will use that app to complete your objectives. Most app

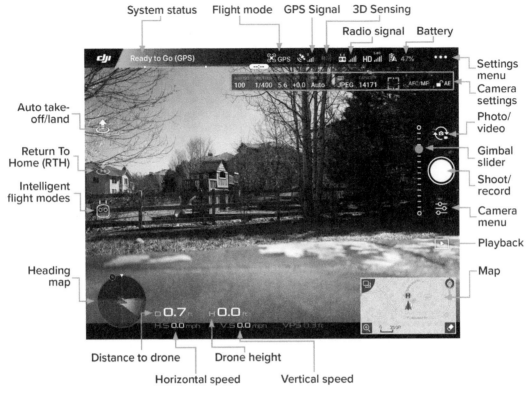

System status   Flight mode   GPS Signal   3D Sensing

Radio signal   Battery

Auto take-
off/land

Return To
Home (RTH)

Intelligent
flight modes

Heading
map

Settings
menu

Camera
settings

Photo/
video

Gimbal
slider

Shoot/
record

Camera
menu

Playback

Map

Distance to drone   Drone height

Horizontal speed   Vertical speed

Fig. 7-2: Screenshot from the DJI GO 4 manual website. The full manual can be viewed at https://store.dji.com/ guides/dji-go-4-manual/ Source: Jon Altschuld.

developers will provide documentation, or even create videos to teach users. There also are a wealth of YouTube videos, forums, and websites where users of these apps share their knowledge and experience.

These resources are not only important when first learning to fly, but also whenever you run into an issue with an app, are curious about its capabilities, or are learning how to accomplish a new objective with the app. For example, Litchi can be used to capture photos or videos for a wide variety of uses. These include capturing 360° photographs, creating waypoint flights with smooth curved transitions, automatically focusing the camera on different points of interest, and creating time-lapse videos. Each of these require an in-depth understanding of the capabilities of Litchi, where controls and settings are stored, and how those controls and settings can be used to arrive at the final product.

Simulator Mode

Some manufacturers, such as DJI, even include a simulator mode in the flight controller app. This allows someone to use the actual hardware (controller, with display device if needed) to practice flying without the drone actually leaving the ground. This can be done inside at any time (day or night), and it helps to train pilots about what their control inputs will cause the drone to do.

For example, most drones have an emergency shutoff procedure. This cuts off power to the motors, even if the drone is in flight. It is a last resort measure, and certainly not something that anybody would want to test while flying their drone. However, it is important to know this procedure in case it is needed. It's also important to know the procedure to make sure it is not done accidentally (most of these require a somewhat complicated and unusual combination of joystick and button input to avoid this). Using the simulator function, a pilot can practice performing an emergency shutoff procedure without ever putting people or the drone at any risk.

The simulator function is also useful for simply learning the basics of flight control, as well as practicing specific flight maneuvers.

Flight Operations

When in the field for flight operations, it is important to have standard practices that all of your team members follow. While each company will have its own set of practices, this section goes over several "rules of thumb" and considerations to take into account when creating flight practice standards.

Flight Planning. Prior to going to the flight area, it is always a good idea to plan out your operation. Often, this begins by viewing the project area in Google Earth with the 3D Terrain option turned on. This bird's eye view of the project area provides context for what is surrounding the site, how you can access the site, where safe operator and flight areas are, and potential issue areas (Figure 7.3).

Fig. 7-3: Planning out the area you can cover with each flight in Google Earth will ensure you are bidding projects appropriately and you are familiar with the site constraints. Source: Chinook Landscape Architecture.

Pre-Flight Checklist. Pre-flight checklists for drone operations are available from many sources online, and they usually include items such as checking propellers, battery level, battery seating, controller battery, memory card, GPS signal, weather, surroundings, closing other apps on the control tablet and putting it into "do not disturb," etc. Having a checklist enforces consistent practice across your entire team, and it encourages a regular procedure for your pilots. This typically results in fewer careless accidents and more attention paid to anything outside of the normal parameters. It is also useful to prove your company has implemented safety protocols in case you do have an accident.

Takeoff

During and directly following takeoff, it is a good idea to always climb to a low altitude (10–15 feet above the ground, for example) and test each of the controls. This means slowly testing that each of the joystick directions and controls operates as expected, and ensuring that all signal notifications (radio signal, GPS, battery percentage, controller battery percentage) are correct and as expected. For example, if you take off and hover, and notice that the drone battery has already dropped by 30%, you know you should land and remove the battery from operations instead of continuing with the mission (Figure 7.4).

Another good rule of thumb for takeoffs is to keep the drone pointed in the same direction as the pilot. Most drone flights are conducted safe distances from possible obstructions, but the takeoff and landing usually pose the most risk because there are inherently obstructions closer to the aircraft. During this time, a pilot's quickest reaction is to move their fingers/hands in the same direction that they are facing. If the drone is facing the pilot, it is much more likely that the pilot will accidently move the drone in the opposite direction than they desire, causing a collision.

Fig. 7-4: Hovering and testing all flight controls at a low altitude at the beginning of each flight is good practice. Source: Jon Altschuld.

During Flight

Flight practice standards during flight will vary by project and objective, but there are some common considerations. These include maintaining flight area awareness, keeping the drone within visual line of sight (VLOS), having an emergency plan, and exercising caution over doubt. It is important for a pilot to know their limitations and their comfort zone, and to not push either. This is also where regular flight training comes in useful; it is better to push your comfort zone a bit on a training flight with low risk (away from obstructions, without the added pressure of billing time, usable deliverables, etc.), than on a project flight with higher risk.

Flight Area Awareness

During drone flights, the perspective of how high and where the drone is in relation to other objects can be very deceiving. This is especially true when trying to determine the height of the drone and nearby objects as the drone gets further away from the pilot. There are two main strategies to combat this. Both require that you take inventory of your flight area and surroundings before taking off.

The first strategy is to utilize all of the information at hand; most flight controller apps provide data on distance from the "home point" (takeoff point), AGL altitude (usually measured from the home point), and even distance to surroundings (on drones with obstacle avoidance sensors).

Let's look at a hypothetical, but very realistic, scenario. You are flying over a treed area and on the live camera feed, the trees look incredibly close to the camera. You know from your pre-flight inventory that the trees are approximately 40–60 feet tall, and they stand on ground that is fairly close in elevation to your home point where you took off. The flight controller app shows the AGL altitude is 90 feet. Even though the trees look very close on the live camera feed, you know from this information that you are safely above them. This is actually very common because the camera on most drones zooms in a certain amount. This causes objects in the live feed to appear closer than they are in real life. Of course, a pilot needs to rely on all available data, including visual line of sight and obstacle avoidance sensors, to fly safely and avoid collisions (Figure 7.5).

The second strategy is to mark the bounds of the project area, and/or interim boundaries, physically on the ground prior to flying. This strategy is less about avoiding collisions, and more about ensuring the correct project area is captured in the dataset. Once in the air, it can be very deceiving where the drone is, and the live camera feed can only provide so much visual information. This is especially true when gathering very high-resolution data because higher resolution typically requires flying closer to the subject, which means seeing less of the overall subject in the live feed, or when flying areas without distinguishing features; the live feed only captures one part of the overall subject at a time.

By marking the subject area, or individual flight areas, with markers that are visible from the flight altitude, a pilot can easily recognize where the drone is in relation to the subject boundaries. These markings can vary from temporary survey paint, to more formal markers, to simply inventorying existing features that are near boundaries. For projects capturing

Fig. 7-5: The camera's live view perspective can be deceiving in terms of how close the drone is to objects. Source: Chinook Landscape Architecture.

rockfaces along highways, big lines and arrows are painted on the roadway and rockface to denote each flight boundary (Figure 7.6). This allows the pilot to know when the drone has travelled far enough to capture the appropriate photographic data for each flight. Prominent features such as sign posts, or uniquely shaped trees, are also used to orient the pilot during flight.

One common approach is to lay out recognizable objects, such as large orange 5 gallon buckets or bucket lids, around a site location. Another option, depending on the site, is to use survey marking paint to mark the extents of each flight, the overall project area, etc. This can help with in-flight navigation and when post-processing data and looking for markers.

Maintaining Visual Line of Site (VLOS)

A challenge that every drone operator will run into is maintaining VLOS. This is required by the FAA but once out in the world there are many reasons operators will lose visual contact with a drone. This can be due to light and cloud conditions, the drone flying behind tree canopies or around corners, or other obstructions. The FAA does note that it is acceptable to lose visual line of sight for short periods depending on site conditions. Here are some suggestions to ensure visual line of site for flying.

Pilot and observer – it is recommended by the FAA to have a pilot and observer. Two pairs of eyes are always better than one. At DHM Design we try and have a pilot and visual observer for most if not all drone flights. For complicated flights it's a given (Figure 7.7).

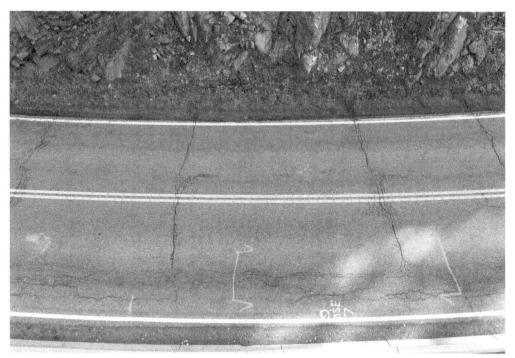

Fig. 7-6: Marking boundaries and overlap areas for flights, as well as locations of obstacles such as overhead power lines will give you a better sense of your surroundings while in flight. Source: Chinook Landscape Architecture and Yeh and Associates.

Fig. 7-7: A visual observer will allow you to focus on flight metrics without losing track of the drone. Source: Chinook Landscape Architecture and THK Associates, Inc.

Hover the drone – if you can't find the drone with a visual scan of the sky, stop and hover. In the case of automated flights, this might be more complicated than during manual flight. You may need to pause the operation to make it hover. After regaining VLOS, you will want to resume the mission; different flight apps have different controls for these procedures. Know all of the features of the flight app you are using.

Frequently check the screen – while you do not want to be using the live camera feed on the control tablet as your primary means of navigation (the FAA specifically states that the live first person view from the camera is not an acceptable substitute for maintaining visual line of sight), it is a useful tool for knowing where the drone is. Most apps allow for the camera to be moved, even during automated flights, so you can scan the area the drone is in for any obstructions if needed.

Pay attention to the drone sensors – many drones now have obstacle avoidance sensors, and some even display the distances to objects. Although the accuracy of these can vary, they are useful for determining the proximity of objects to the drone and maintaining visual line of sight. In many instances the drone sensor will trigger an alarm indicating its getting close to an object.

Bring binoculars for the visual observer – this is also recommended by the FAA and is an effective measure to locate the drone in the sky.

Emergency Operations

It is important to consider emergency actions during flights and have set procedures in place before encountering these emergencies. These include what to do if the controller and drone lose radio connection, if the drone has a "fly-away," if a manned aircraft is approaching, if the battery is suddenly too low to return to the home point, etc. Having a set procedure in place for each of these will reduce the risk for injury, property damage, and as a result, legal action.

Most drones have a "return to home" button/feature; if for whatever reason you need the drone to come back, you can press the return to home button and the drone will return to the original launch point (hopefully near you). Many apps, like DJI Go, allow the return to home button settings to be programmed with specific parameters such as rising to a specified altitude before flying home to avoid any potential objects.

Most flight controller apps also allow programming for certain unexpected situations such as a lost signal. With this programming, the drone can be told to automatically enact the return to home protocol, hover in place, etc. if the signal between controller and drone is completely lost.

For automated flying, the ability to take control of the drone for an emergency or other reason is also typically straightforward. This can depend on the specific drone and app being used to fly, but you should always understand how an app allows you to retake control of the drone and learn about the manufacturer's recommended controls prior to any flight, takeoff, and landing.

It is a necessity to think about and prepare for emergencies when flying. Like most accidents, they can happen quick and no human or even automated reaction time can course correct.

The last rule of thumb for flight operations is: "when in doubt, just hover." Just about every drone that would be used for professional practice is controlled not only by the pilot, but also by GPS positioning technology. This is why these drones will hover and hold in place, even when battling winds. Letting go of the controls forces the drone to maintain its current position. Although an exit plan is important when flying close to objects and something unexpected happens, it is just as important to not rush a decision if you need to think about the best solution. If you have doubt about the best way to proceed or need to think about the next step, just hover in place and take the time to arrive at the best solution.

Landing and Post-flight

As with taking off, landing with the drone facing the same direction as the pilot helps to avoid control errors. After landing, it is a good idea to have either a post-flight checklist or at least a post-flight procedure that accounts for examining the equipment and properly logging the flight. Common items to inspect and record over time include battery trends, propeller damage/condition, gimbal movement, and controller input reaction time.

Automated Flying

The more you fly and do flight missions the more you will utilize both manual and automated flying. For many projects, automated flights can capture more consistent photograph angles and overlaps, as well as smoother video paths. The paths can also be saved to be re-flown at future dates and compare site conditions over time.

Many flights, particularly nadir photogrammetry flights (see Chapter 9 for more details), and certain video capture techniques, utilize flight controller apps that automate the flight operations. These apps require some pre-flight inputs such as flight area, route (grid) spacing, altitude, etc., and then the app will control the takeoff, fly the route, capture photos, and land the drone. Although this may sound like the easiest way to fly a drone, we are discussing it after looking at best practices for manual flight for a very important reason: whether automated or manual, *it is still the pilot's responsibility to be in control of the aircraft at all times*.

Pilot Control

Maintaining the ability to be in control of the aircraft during automated flights means monitoring the flight and knowing how to pause the operation and take control of the aircraft at any point. In order to do this effectively, the pilot must be familiar with everything relating to manual flight controls and standards, as well as the controls and functionality of the automated flight controller app.

Manual Intervention

It is not uncommon for automated flights to require manual intervention for a variety of reasons. The operation may be flying too close for comfort to nearby structures. For nadir photogrammetry flights, some apps will automatically slow the drone's speed to account for lighting conditions. As a result, an operation planned for minutes can be changed to hours because of

Fig. 7-8: Screenshot of the Litchi Mission Hub Planner in Chrome. Through the Mission Hub, you can plan waypoints and flights from your office, saving field time and increasing your site awareness. Source: Chinook Landscape Architecture.

dim lighting or harsh shadows. A smooth curved video flight may be travelling too fast for the desired video. An operation may be taking place within controlled airspace (with FAA permission) and a manned flight is seen entering the operation area. In all of these situations, it is crucial that the pilot know how to intervene and take control of the aircraft.

Pre-flight Planning for Automated Flights

One key advantage of automated flights is the ability to pre-flight program the automated flight paths. This is usually done on a laptop or desktop, although many apps require flight planning to occur on the smart device or controller being used with the drone.

Regardless, pre-flight automated flight planning is the ideal way to ensure safe flights and allows pilots to know a location before they even arrive, which is especially useful for new or unfamiliar locations. Pilots will have to learn to become proficient with these pre-flight tools. In many instances, you will spend more time doing automated pre-flight paths than the actual flying of the drone (Figure 7.8).

Acquiring and Working with Drone Data

CHAPTER 8

Imagery and Videos

I magery and video are at the heart of drone data. They provide unprecedented views and quick visual analysis. Capturing smooth, professional quality video is equally valuable when it comes to analysis and gaining a higher perspective of a site. Similarly, detail rich graphic renderings are easily generated using annotation, photo editing, and 3D modeling software in combination with drone imagery and videos. This chapter explores the different ways imagery and videos can be captured and used to create infographics and renders (Figures 8.1 and 8.2).

Photo and Video Quality

At the time of this writing, most professional drones come equipped with cameras that can collect 4K images and video (3840 × 2160 to 4096 x 2160 pixels). Of course, a more customized drone setup can be configured with just about any standard DSLR camera. In the future, higher quality outputs are sure to be available (Figure 8.3).

4K images and videos offer a superior level of detail, great for the type of visual work described below. For most professional work, shooting the highest quality resolution is recommended because it will provide the best results and flexibility when working with images in editing software.

Using DJI Standard Apps

The DJI standard flying apps come loaded with camera and recording features. At its simplest, the app allows you to adjust the camera and capture photos, or to start a video recording, by clicking on the button in the app. It includes the standard ability to adjust resolution or image type (RAW versus JPEG) directly through the app. The apps provide the ability for more

Fig. 8-1: Bird's eye drone image of Hudson Gardens in Littleton, Colorado. The base image is used for 3D visualization. Source: DHM Design.

Fig. 8-2: The resulting image, post-processed with Photoshop, integrating a SketchUp and Lumion render creating an expressive, context rich image. Source: DHM Design.

Fig. 8-3: Drone capture image of a healing garden Labyrinth in Durango, Colorado. Chris Fortin and Justin Clark capturing marketing videos for a built DHM Design Project. Source: DHM Design.

advanced uses, including the ability to follow a tagged object (like a person walking, car, or biker) or add a focal point for videos.

Image Geotag

Photos and video captured with a drone can be geotagged (it requires a drone connected to GPS), meaning they have metadata attached to the image that include the location, height, camera information, and much more (Figure 8.4) This is important information for the drone to capture as it is required for photogrammetry software to create overlaps and process the data. It also provides the ability to create geotag tours in Instagram and other apps.

Litchi Flight Planning Software

The Litchi Platform is mentioned in Chapter 5 and is a must have app for recording videos, particularly with smooth and complex flight paths (https://flylitchi.com/). This simple but powerful app is the way to capture smooth, professional looking video records by being able to create pre-programmed recording flight paths. This can be done in the field (discussed in the Litchi Smart Device App section below), or it can be done in the office. In the office, flight planning can be done on the smart device that will control the drone, or it can be done online through the Litchi web portal (Figure 8.5).

What Litchi does so well is allow you to pre-program each and every step of a flight using waypoints. Each waypoint can have a custom flight speed, drone elevation and direction, and angle the camera is pointing. It is possible to set focal points for each waypoint and issue

Fig. 8-4: To access the Geotag information on a Windows computer, right click on the image, select Properties. From the properties menu select Details. Scroll down to GPS. Source: Image by Daniel Tal.

commands when the drone hits a new waypoint like stop recording, take a photo, stop, rotate, etc. Litchi also allows you to tell the drone to focus the camera on points of interest (POIs). For example, you can program the drone to focus the camera on POI #1 for waypoints 1–4, and then focus the camera on POI #2 for waypoints 5–7. This, combined with Litchi's ability to interpolate smooth flight curves between waypoints, enables you to capture videos that smoothly transition between both waypoints and camera focus points. Trying to achieve the same results with manual flight control is almost impossible or requires painstaking discipline and reshoots.

Another feature offered by the app is the ability to "hug" the terrain; the drone will adjust flight height based on the elevation underneath it. However, this height is based on distance above ground at waypoint 1. This does work well but should be watched where there are extreme terrain changes as this can cause the drone to drop or behave erratically. Please note: Litchi is basing its height information based on the GPS elevation provided at waypoint 1.

Litchi also allows for the continuous capture of photos at designated intervals. Instead of recording video, the app can be set to collect image burst at intervals from two seconds or more during flight. Other options related to photos include setting the photo resolution, aspect, and

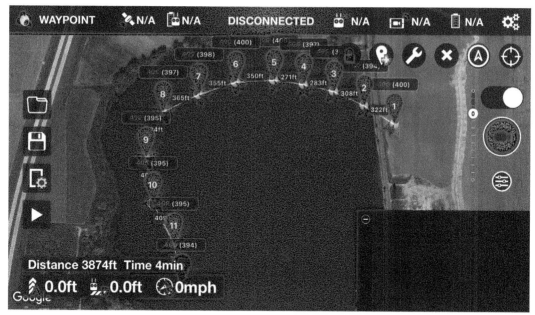

Fig. 8-5: The Litchi App for a smart phone or tablet is easy to use, full of features and allows for excellent quality video and photo capture. It includes the ability to plan missions in the field. Source: Image by Daniel Tal.

format type (JPEG versus RAW). This feature is a great way to collect high quality photos from a flight path instead of having to extract images from a flown video recording which causes reduction in quality.

Litchi Smart Device App

The app is intended to be installed on the device being used to fly the drone: smart phone, tablet, or drone controller. Once installed, log into the app and you can access, load, copy, start, modify, and delete missions. All of it is stored in the Litchi Cloud. Whatever device you log in with that has the Litchi app will have access to all your missions.

Once a mission is uploaded, modifying the waypoints is possible and fairly straightforward. While not as easy as using the desktop browser, you can make changes out in the field or create new missions.

Creating missions in the field with the smart device app does have many useful applications. For example, you may want to create a mission that starts and/or ends with specific views. Creating a mission in the office through the desktop web portal does not allow you to do this. Alternatively, in the field you can fly to the desired starting location, aim the camera to the view you want, and save the waypoint. You can continue this process until all waypoints are saved. Next, you can save the mission and the drone will fly the smooth, interpolated path.

The app does also allow for camera adjustments while flying in mission simply by using the standard camera controls. It is not uncommon to adjust the camera angle at the start of the mission to ensure the best viewing angle.

Litchi Desktop Web Portal

While you can create a new mission through the app on a smart device, Litchi also has a free interface accessed through a web browser. This is the best way to create new missions and flight paths; it's generally easier to use the tools on a desktop or laptop than on a smart device (Figure 8.6).

Creating missions can take minutes, then can be saved and accessed on your smart device used to control the drone. Once you are at the mission location, open the app, load the mission, upload it to the drone, and it will fly the mission as programmed. If capturing video, make sure to start recording video before telling the drone to begin the mission.

Take the time and create thorough missions. The more time spent, the better the result and the easier it is to fly a site. Using the desktop interface also allows for greater familiarity with the site and all of Litchi's features.

Virtual Litchi Mission

There is a third-party app for Litchi called Virtual Litchi Mission or VLM. An exported Litchi mission (exported as a CSV file) from the desktop browser can be imported into VLM and then exported as a KML file that can then be viewed in Google Earth (Figure 8.7). Turning on the 3D terrain and objects in Google Earth, you can then "fly" the mission virtually in Google Earth, then assess and adjust the flight path as needed back in the Litchi apps and browser (you cannot modify the mission in Google Earth). It's the best way to preview a mission without actually flying, but it is important to recognize that the Google Earth terrain is not an exact representation of what is currently present on-site (Figure 8.8).

Fig. 8-6: Litchi Desktop app is an excellent tool to develop and plan out flights prior to going on-site. Users can easily upload the mission to the Litchi app to fly, export or import other flight paths, and keep track of past missions. Source: Image by Daniel Tal.

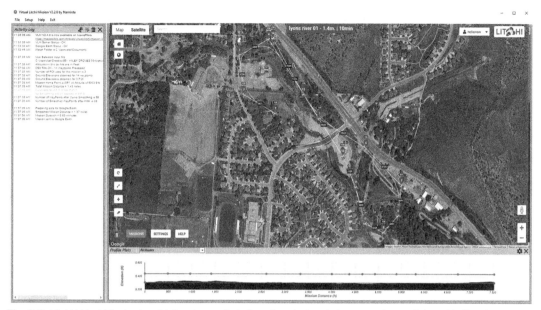

Fig. 8-7: Litchi Virtual Mission is a third-party app that allows for missions to be opened up and viewed in Google Earth. Source: Image by Daniel Tal.

Fig. 8-8: Litchi Virtual Mission shown in Google Earth showing the drone flight path, the direction of the camera, and the height of the drone. Users can play the flight path as a video to simulate what will be captured by the drone. Source: Construction image of Lyons Colorado Drainage way. Image Courtesy of DHM Design.

Working with Drone Photos

Overlaying information on photos has been around since almost the start of photography. Drone photos allow for another permutation of superimposed information. Before drones, balloons, planes, and satellites provided imagery as bases for infographics and visual simulations.

However, drones have given the average professional easy and quick access to aerial imagery that can be used as the basis for these graphics. Possibly even more important, the imagery collected from drones can be customized to individual projects, as opposed to using what is available via satellites or planes. Below, three common ways of creating visualizations from drone imagery are discussed.

Annotated Images

The simplest way to utilize drone imagery in professional practice is to "mark up," or annotate images (Figure 8.9). Typically, these are simple annotations such as text, labels, and simple 2D overlays placed on top of drone images. This can be done in several different software packages, including Adobe InDesign or Photoshop. For final presentation, these can be shown by themselves, on boards, within reports, in construction documents, or within PowerPoint presentations. These are also easy to combine with drone videos: a drone video plays and then pauses on the overlaid images. This is easy to accomplish with most presentation software like PowerPoint or similar.

Fig. 8-9: Annotated image of North Fruita Data showing sustainable biking trail improvements. Source: Chinook Landscape Architecture and THK Associates, Inc.

Photo Matches

A step up from annotated images in terms of complexity, a photo match entails "matching" drawings of proposed features to a particular drone photo. The "matching" requires paying close attention to the scale, angle, and focal length of the drone camera. The simplest type of photo match is a plan view, and if the drone imagery has been processed with photogrammetry software (see Chapter 9), proposed features can be drawn to scale over the image (Figure 8.10).

Generally, the proposed features of a photo match are created either by hand sketch, in Photoshop, or in a 3D modeling software such as SketchUp (Figures 8.11 and 8.12).

Hand Sketch Photo Matches

Hand drawn plans, overlays, and perspectives can easily be placed over images. Depending on the project and view, as well as the artist, it is often easier to draw on tracing paper directly over the drone photo. This is essentially the same process as drawing a traditional plan or perspective view, except that the drawing is being created on top of an image base. Final alignment and compilation are often done in photo editing software such as Photoshop. (Figures 8.13–8.16).

Photoshop Photo Matches

This is essentially the same process as the hand sketch method described above, except that instead of hand drawing proposed features, a photo editing software such as Photoshop is used. Likewise, instead of drawing the proposed features, the software is used to compile the final image. There are several different techniques to accomplish this. As with the hand sketch

Fig. 8-10: Existing conditions image of Little Dry Creek in Denver with the Denver skyline in the background. Source: DHM Design and City of Englewood, CO.

Fig. 8-11: A SketchUp model and Lumion render are integrated into the drone image to create an expressive photo-match graphic. Source: DHM Design and City of Englewood, CO.

Fig. 8-12: Composite image showing the existing conditions faded out to highlight the proposed design for Little Dry Creek. Source: DHM Design and City of Englewood, CO.

Fig. 8-13: Aerial image of I-90 Interchange in Billings, Montana. 2D CAD line work is overlaid on to the aerial. This is the first step in setting up the photomatch. Source: DHM Design.

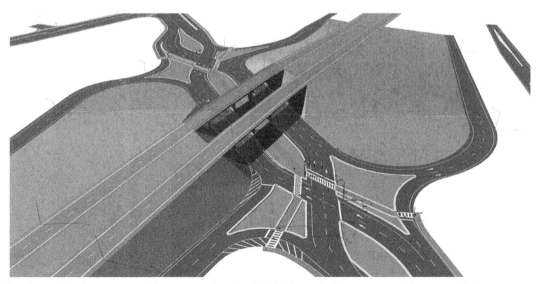

Fig. 8-14: A 3D SketchUp model is generated from the 2D CAD line work. The manual modeling process is the most time intensive task. However, by not needing to develop 3D context, a considerable amount of time is saved. Source: DHM Design.

Fig. 8-15: The bird's eye drone image of existing conditions is selected for the model photo matching.
Source: DHM Design.

Fig. 8-16: The final image is completed in Photoshop. Perfectly merging the model with the drone image.
Source: DHM Desgin.

method, Photoshop compositing is a method that has been used for many years, but adding drone imagery as a base gives it much more context (Figures 8.17 and 8.18).

3D Model Photo Matches

Although these are more complex to produce and require the use of 3D modeling software, 3D model photo matches are more accurate than the previous two methods. These are great for architectural and urban visualization. They are the best way to illustrate before and after conditions in exact context.

Various 3D modeling programs can be used to create the model: SketchUp, 3ds Max, Rhino, Revit, Lumion, Maya, etc. in conjunction with Adobe Photoshop or similar 3D editing software (Figure 8.19).

Like the other photo match strategies, the 3D model is merged with the drone photo. This requires the painstaking process of lining up the 3D model with the selected image or images (camera location, camera angle, focal length, etc.).

One of the most challenging parts of this process can be attempting to line up images and 3D models that are spread out over large areas. For example, a single building in a relatively close drone bird's eye image is easier to compile then a large spread out location, like a camp site or a building complex. Even for experienced photo editors, the process can take a little bit of time to master (Figures 8.19–8.21).

Fig. 8-17: The raw drone photo shows the existing conditions as the site is under construction. Source: Chinook Landscape Architecture.

Fig. 8-18: Post-production in Photoshop allows for annotation and expressive photo improvements including trees, water, and grasses and showing the completed home. Source: Chinook Landscape Architecture, Brandon Parsons, and Wright Water Engineers.

Fig. 8-19: SketchUp model imported into Lumion showing the site and trail improvements. Source: DHM Design.

Fig. 8-20: The drone image is overlaid over the model. The model, in Lumion, is rotated and adjusted to match the photo. Lumion includes an image overlay feature making it easier to get the model view aligned to the image. Source: DHM Design.

Fig. 8-21: The merged images show the site concept in the context of the real location. Source: DHM Design.

Some pro tips:

1. First start by geolocating the model on a 3D aerial if available. This can assist in aligning the 3D model with the bird's eye image from the drone.

2. Complete the modeling of the proposed features.

3. Line up the 3D model with the image. There are various ways of doing this – Peek Through (free desktop app) allows you to make one software panel transparent. In this case, the SketchUp window is made transparent showing the drone image underneath. The model is then aligned, as best as possible.

   Similarly, the 3D software Lumion allows for a base image to be inserted and then made transparent to line up the Lumion model to the drone image.

4. Try and get it close. It can be challenging, even with programs like 3ds Max, which can help automate the process, to get 3D modeling software to exactly align with the photograph. One strategy is to create simple models of existing features that show up in the photo and that you know dimensions of. For example, if there are 70 feet tall power poles in the photo, model 70 feet tall structures in the correct locations within the 3D model. These will serve as "benchmarks" for you to align the 3D model view to the photo with.

5. Export an image from the modeling software. Ensure you only export what you need and try to include a "green screen" background in the export too.

6. In Photoshop combine the drone image with the 3D model. This can be tricky. There are a couple of useful methods. First, mask out the renderings' artifacts or locations that are not needed. You just want the objects/models that are the focus of the render.

   After masking is completed, use the Free Transform tools while making the layer transparent and tweak the image to best fit with the image. This might require stretching the model image into place.

7. If needed, go back to the model and adjust the view to get to a better fit. Keep the photo editing software file with all its masks and layers. This way a new render can be inserted and have the mask applied without having to redo the work beyond some simple tweaks.

   Another suggested tip is to use clone stamp to copy and paste aspects from the photo over 3D modeling elements. For example, use vegetation from the photo or other sources instead of 3D modeled trees. This can be done with cars, ground plane, and other elements to add to the overall realism of the image.

Working with Drone Videos

There is a growing desire for merging 3D animations and drone videos. These compelling multi-media videos show an existing real condition (drone footage) that is then overlaid with a 3D model, just like the drone photo match, except the model is animated and fits neatly within the drone footage. Although this sounds like a simple concept at first (especially if you are already doing drone photo matches), matching a 3D model into a moving video is much more complex.

There are many ways to achieve this but they require more specialized knowledge and understanding of modeling, rendering, and video editing software. At minimum, it requires an understanding of the following technology and methods. Suggested software is included below but by no means is this the final word on how to achieve merged animation and drone video footage.

▶ *Camera Tracking.* Camera tracking refers to the 3D path the drone took to capture the footage (Figure 8.22). The "track" needs to be derived from the footage so it can then be used in the 3D model animation. The track is then inserted into the 3D modeling software where it is used to create an identical path for the 3D model. The animation then matches the drone camera track and the two can be overlaid on each other, matching track for track.

Adobe After Effects allows for the import of a footage, like a drone video, and then the extraction of the camera track. This process does require some time for the software to complete and it helps to have a robust computer (see the section on hardware in Chapter 5). Once the camera track is processed and saved, it will be used again in After Effects once the 3D model animation is completed.

Fig. 8-22: Camera tracking allows for the creation of animated video overlays merged with drone recorded video. Source: Image by Daniel Tal.

- *3D Modeling Software.* Several 3D modeling software packages allow for the import of a camera track created in Adobe After Effects or similar software. Maya, 3ds Max and Blender are three examples. All three programs require experience and good modeling skills to use. When building such a model, like with photo matching above, it is important to geolocate the model. This will help with matching the camera track to the model versus the location it needs to be merged into. Again, having an aerial base underlay to model over will assist in this process.

  Next, the 3D model is generated in one of these software packages or similar. This means generating the geometry, textures, and features of the model. Once complete, the imported camera track is used to develop an exact animation path. The resulting animation should match the speed, height, camera direction, and aperture of the drone footage.

- *Merging Footage with Model.* The last step is importing the 3D model back into After Effects (or similar software) that includes the original extracted camera track. The animation is overlaid to the camera track and the drone footage in the movie time line. From there, the user must "cut out" and align the 3D model animation with the footage. Once completed, the video can be exported for presentation.

Pix4D (and Other) Animations

From a photogrammetry model or point cloud, most software like Pix4D, has features to create flight path animations from that data.

Photogrammetry

In this chapter, you will learn how drones can be used to create 3D data and geometry from photographs with a process called photogrammetry. This process is fairly simple with the technology available today, but capturing good data requires careful planning and an understanding of how the technology utilizes photographs. While the information in this chapter can be applied to many drone and camera combinations and software packages, it adheres most closely with DJI Phantom and Mavic drones and Pix4D photogrammetry software.

The photogrammetry process yields a variety of data outputs. They include:

- ▶ Classified or unclassified colored point clouds
- ▶ 3D textured meshes
- ▶ Orthorectified imagery
- ▶ Digital elevation models (DEMs)
- ▶ Topographic contours
- ▶ Reflectance and index maps

What Is Photogrammetry?

According to Merriam-Webster, photogrammetry is "the science of making reliable measurements by the use of photographs and especially aerial photographs" (https://www.merriam-webster.com/dictionary/photogrammetry). In the photogrammetry process this chapter discusses, those measurements are used to create 3D data and geometry in the form of point clouds and meshes (Figure 9.1).

Fig. 9-1: Drone photogrammetry data along I-70 in Colorado. From left to right, initial point cloud, final point cloud, 3D mesh. Source: Chinook Landscape Architecture and HDR, Inc.

Photogrammetry requires a series of overlapping geotagged photographs. The geotag provides information about where each photograph was taken from. Photogrammetry software uses this information, as well as specific details of the camera such as sensor size, focal length, and shutter type to determine points within the photographs. The software identifies points that are present in multiple photos, and uses geometric algorithms to identify three coordinates for every point: X, Y, and Z. Most photogrammetry software also identifies a red (R), green (G), and blue (B) value based on the colors in the photographs. By the end of this initial processing, the software will have created a 3D point cloud (Figure 9.2), in which every point has six values – X, Y, Z (location), R, G, B (point color). When viewed from afar, the point cloud often looks like a solid 3D mesh object. In reality, what is produced are individual points located in 3D space forming the point cloud and the basis of the 3D data. The information can be further refined, classified, and processed into a 3D mesh, which is ideal for additional 3D modeling in software like SketchUp, for example.

Drones and Photogrammetry

Although photogrammetry has recently gained a lot of support and new software packages, the process has been around for decades. The widespread use and affordability of drones combined with high quality digital cameras has brought photogrammetry to the masses. Sites and projects that previously would have required custom, high quality, expensive airplane or satellite acquired imagery can now capture imagery much quicker and cheaper, and often at a higher resolution.

Fig. 9-2: Drone photogrammetry point cloud. Each point has six values – X, Y, Z, R, G, B.
Source: Chinook Landscape Architecture.

Drones equipped with GPS receivers and camera equipment can be used for photogrammetry. The drone must be able to geotag the photographs with the GPS data and camera metadata (Figure 9.3). Based on the site and the type of project, the drone can be flown manually, or it can be flown autonomously through flight programming apps discussed later in this chapter. Because most of the consumer and professional grade drones available today have these as standard features, drone photogrammetry has become common in the architecture, engineering, and construction industries.

As a result of this booming use, several software packages have been developed and quickly advanced to meet user requests and industry demands. Drone photogrammetry has become common for surveying, site inspection, site analysis, construction monitoring, and for creating base 3D models of existing conditions for multiple uses. The 3D base model can be used for various levels of design (based on the accuracy and precision of the model), for site analysis by specialized trades (such as geotechnical engineering), or for visualization of proposed designs.

Photogrammetry Accuracy and Precision

The accuracy of photogrammetry results is at the heart of many discussions on its applicability in professional settings. In order to understand photogrammetry accuracy, one must re-consider how they think about accuracy. Generally, there are two types of accuracy to consider:

1. Relative accuracy – how accurate are points and measurements within the project site.

2. Absolute accuracy – how accurate are the coordinates of points in the dataset in reference to a larger coordinate system.

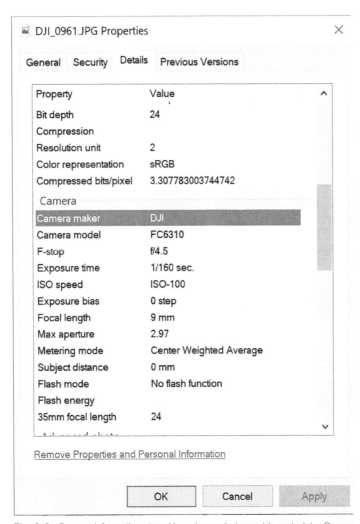

Fig. 9-3: Camera information stored in a drone photograph's metadata. Source: Chinook Landscape Architecture.

It is also important to realize that the accuracy on a photogrammetry project is not a single uniform number. While there are certain metrics to gauge a project's average accuracy, this can change with a number of variables. This is because photogrammetry data is based on photographic data. If the photoset does not capture enough of a specific feature, or if a feature (such as vegetation) is moving or objects are in different positions, this can impact accuracy. In general, the accuracy (both relative and absolute) of a photogrammetry dataset will depend on:

▶ Camera sensor and distance to subject – if the same site is flown at the same altitude with a low-resolution camera and a high-resolution camera, the high-resolution photoset will yield higher accuracy results. Similarly, if the same camera is used to fly the same site, but at different heights (distances to the subject), the dataset flown lower/closer to the subject will yield higher accuracy results (Figures 9.4 and 9.5).

Fig. 9-4: Zoomed in images of the same subject taken from different heights. The above image was taken at a lower altitude (closer to the subject) and is a higher resolution, meaning each pixel represents a smaller area in real life. Source: Chinook Landscape Architecture.

Fig. 9-5: Zoomed in images of the same subject taken from different heights. The image above was taken at a higher altitude (further from the subject), and is a lower resolution, meaning each pixel represents a larger area in real life. Source: Chinook Landscape Architecture.

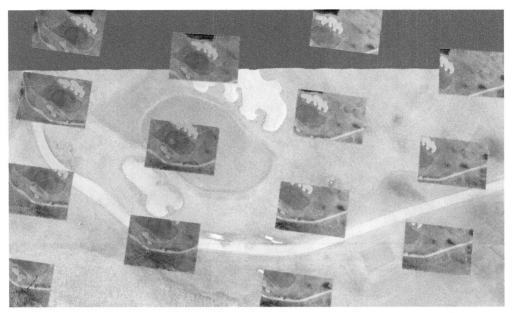

Fig. 9-6: In this Pix4D screenshot, each thumbnail shows an individual drone photograph at the approximate location taken. The final photogrammetry 3D mesh is shown faded in the background. As you can see, there is plenty of overlap (approximately 80% in this case) between the photographs. Source: Chinook Landscape Architecture and ECI Site Construction Management, Inc.

Fig. 9-7: This project site was relatively flat, but it also contained vertical walls. In the screenshot, you can see nadir images were taken looking straight down (roughly normal to the ground), as well as images taken at a slightly downward angle looking at the wall (roughly normal to the wall). Source: Chinook Landscape Architecture and ECI Site Construction Management, Inc.

- Photograph overlap – similar to the last point, it is important to remember that photogrammetry results are based on photographic data. One of the most common reasons for poor photogrammetry results is a lack of overlap (Figure 9.6).

- Image quality consistency – over-exposed, under-exposed, blurry, etc. photographs will negatively impact the project's accuracy.

- Flight path normality – keeping the drone camera at an angle roughly normal (perpendicular) to the subject is important in collecting good data, especially for complex subjects. This is why capturing only nadir (top down) images of a vertical structure does not produce good data (Figure 9.7).

Ground Control Points

In order to create an accurate and precise photogrammetric model, ground control points (GCPs) are required. A GCP is a point within the project that has known coordinates within a specified coordinate system. For photogrammetry purposes, that point must be visible in the photographs. During processing, the known coordinates of these GCPs are added to the dataset and the entire model is rectified to be correctly geolocated and have a higher relative accuracy.

Although the drone will record GPS data for every photograph, that data is only so accurate. GPS receivers are fairly accurate horizontally (latitude and longitude), but they are notoriously inaccurate vertically (altitude/elevation). In the authors' experience, the drone GPS is typically within 1–3 meters accuracy compared to known points. Vertically, however, that accuracy can easily grow to 20–50+ meters, especially when in areas where GPS signals are sparse.

Adding accurate ground control points is required for achieving reasonable absolute accuracy, and it also improves the relative accuracy of a project. In fact, the accuracy of the GCPs is typically the limiting factor for the absolute accuracy of a photogrammetry project (Figure 9.8).

There have been numerous studies and white papers on the accuracy of photogrammetry in specific fields. Many of these cover photogrammetry in a more technical fashion, considering things such as base-height ratio, wide versus short baseline pairs, etc., in addition to just overlap, distance/resolution, image quality, and GCPs. The collection and marking of GCPs are discussed in more detail in the section below: GCP planning.

Also, below are links to some of these white papers/studies:

https://s3.amazonaws.com/mics.pix4d.com/KB/documents/Pix4D+White+paper_How+accurate+are+UAV+surveying+methods.pdf

https://www.pix4d.com/blog/getting-expected-accuracy-pix4dmapper

https://www.mdpi.com/1424-8220/15/3/5609

http://www.c-astral.com/media/uploads/file/Bramor%20Accuracy%20compare_RTK_GPS.pdf

Fig. 9-8: GCPs shown in Pix4D. Spreading GCPs out across a project site improves the overall accuracy throughout the site. Source: Chinook Landscape Architecture and ECI Site Construction Management, Inc.

Collecting Data

The first step to any drone photogrammetric project is collecting the data. In order to collect usable data, pre-flight planning is necessary. Depending on the type and location of the project, this planning typically includes flight planning, image collection planning, weather planning, ground control points planning, and other project-specific planning requirements.

Flight Planning. Prior to going on-site, it is very important to create a flight plan that outlines the individual flights of the operation. This flight plan needs to define the approximate extent and location of each flight based on numerous criteria. These include flight (battery) time, height of subject(s) to be inventoried, and the shape of these subjects. The flight plan also needs to account for FAA rules (or the rules of the governing agency) such as keeping the drone within VLOS, not flying over people, not flying at night, checking airspace requirements, etc. An easy way to document and present this flight plan is to create named polygons in Google Earth for each individual flight. Text can be added to the polygon notes in Google Earth, or a separate document can be created with notes on each polygon.

Although some flight control apps allow you to plan and route individual flights on-site, advanced office flight planning is much more complete. The flight plan is important for flying safely, but it is also a crucial first step to ensuring any drone photogrammetry operation is completed on time and within budget (Figure 9.9).

Image Collection Planning. There are two predominant approaches to collecting drone photographs for photogrammetry – nadir/automatic, and oblique/manual.

Fig. 9-9: Flight planning can be done in Google Earth and is an important first step to properly pricing, scoping, and completing drone operations. Source: Chinook Landscape Architecture.

Nadir/Automatic. Nadir simply means that the camera is pointing directly down when taking photographs. In this scenario, the drone flies a regular grid above the project area and captures photographs at a set interval. The grid is typically flown at a consistent above ground level (AGL) altitude. This altitude will correspond directly to the resolution of the photographs taken, and in turn, the ground sampling distance (GSD) of the photogrammetric data. Just as the resolution of aerial photographs is typically defined as the pixel resolution (i.e. 1 pixel = 1 inch), the GSD defines the distance between the center of 2 pixels in the photogrammetric point cloud. A lower value GSD means a higher resolution result. The higher the altitude of the flight, the greater the GSD, and the lower resolution of the final result. Another way to think of this is: the further away the drone is from the subject (the ground in this case), the lower the resolution of the images, and the lower the resolution of the final result.

Nadir flight grids are often flown with flight control apps like Pix4Dcapture or Maps Made Easy Map Pilot. These apps allow you to pre-program a grid for an area based on the amount of overlap (front and side) and altitude of the flight. During flight, the app controls the drone and captures photographs, but the pilot does have the ability to pause the operation or take control at any time (Figures 9.10 and 9.11).

Nadir flights typically require less experience and piloting skill because the drone is flown far above any structures or landforms, and the photographs are taken automatically by the flight control app. Generally, photographs should have at least 70–75% overlap, with more overlap for heavily vegetated areas or more complex terrains.

Fig. 9-10: Example of a nadir image taken via drone. Source: Chinook Landscape Architecture.

However, nadir grid flights will not work well for very complex terrains, structures, or vertical surfaces. Because the photographs are only being taken from one angle, the entirety of these features are not fully captured. Additional flights can be flown along the same grid, but with the camera rotated to a non-nadir angle. The flight grid can also be rotated 90° on additional flights. Both of these approaches will improve the quality of the photogrammetric process, although it still isn't the best solution for complex subjects that have vertical, near-vertical, and/or overhanging features.

Oblique/Manual. The other way of capturing drone photographs for photogrammetry is by taking oblique photos through manual flights. This can be necessary when flying in tight or constrained areas, or when the subject is very complex with vertical and overhanging surfaces. Rather than pre-programming a grid that automatically performs the flight and captures photographs, the pilot manually flies the drone and captures photographs. This requires more planning, and typically more pilot experience, in order to collect a usable set of photographs for the photogrammetry processing.

A perfect example of this is using drones to collect 3D data of rockfaces along highways through the Colorado Rocky Mountains. Author Jon Altschuld and Chinook Landscape Architecture have completed multiple operations to this end. For these operations, a curved vertical grid was manually flown. A good way to think of this is to first consider the horizontal flight grid used in the nadir/automatic scenario. The main subject, the ground, is roughly parallel to the grid. In the rockface scenario, the main subject is irregular and near vertical. Accordingly, consider taking the horizontal grid, turning it on its side, and wrapping it around the general curve of the rockface to be roughly parallel to it. Rather than the drone flying over

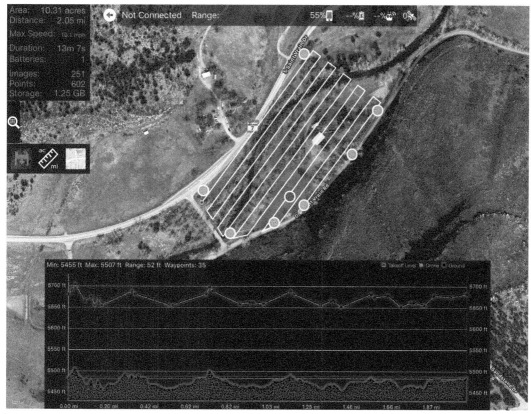

Fig. 9-11: Screenshot of a horizontal nadir flight pattern, routed in the Maps Made Easy Map Pilot for DJI app. Source: Chinook Landscape Architecture.

the subject, it flies in front of it, making passes at varying altitudes and camera angles (Figures 9.12 and 9.13).

Unfortunately, currently there is no good solution to automate these flight paths. This is in large part due to the confined spaces and irregular subjects. These flights are flown manually by the pilot, and the pilot also manually captures each photograph through the drone. The pilot must be very aware of the surroundings in order to avoid collisions with vegetation, structures, and the terrain. They also must be aware of how the terrain and subjects vary in order to get more overlap for more complex features such as vertical faces, overhangs, or vegetation. Because these subjects are typically more complex, more photograph overlap is needed (at least 80%).

Combining Collection Methods. In certain cases, it can be very productive to capture photographs with both methods described above. Let us consider a project site that is generally open but has a main house as the primary feature. It would make sense to complete a nadir/automatic flight for the whole site, as well as two or three oblique flights that fly a circle around the house. Each oblique flight would be flown at a different altitude. For this scenario, there are even some flight apps (Litchi) that can automate the flight path and photograph taking for

Fig. 9-12: Example of an oblique image taken via drone. Source: Chinook Landscape Architecture and Yeh and Associates.

Fig. 9-13: Oblique/manual flight paths viewed in Google Earth. Source: Chinook Landscape Architecture.

the oblique circular flights. If there are other structures nearby that may interfere with these flights, they should be flown manually in the interest of safety.

When considering flight path planning, remember that good data will require the drone camera to be roughly normal (perpendicular) to the subject. We call this the "normality rule." Consider a project site with a house, for example. To capture images roughly normal to the roof and the surrounding site, a horizontal flight grid capturing nadir images would be good. However, to capture images that are roughly normal to the sides of the house, the drone will need to fly around the house taking pictures looking horizontally (or at a slight angle downwards), most likely at different elevations.

Weather Planning. As with any drone operation, weather planning is an important step. From a safety standpoint, it is important to make sure that the current weather and the weather forecast support safe flying conditions. From a technical standpoint, photogrammetry requires a very large number of very clear and well-exposed photographs. Not only does this require a substantial amount of flight time in stable flying conditions, but it also requires the subject be as evenly lit as possible during the flights. If using an automatic flight app such as Map Pilot, there is usually an option for the app to automatically slow the drone based on lighting conditions. This means that the drone will slow down substantially if the subject is in dark shadow. This allows the camera to take properly exposed photographs at a slower shutter speed. If the drone is moving quickly while the camera is using a slower shutter speed, the photographs will be blurred. For mountainous areas or areas with very dark shadows during part of the day, this can change the time of an operation from minutes to hours (Figure 9.14).

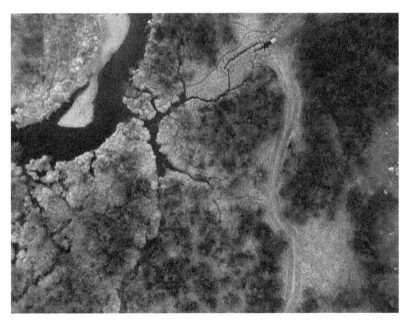

Fig. 9-14: Flying a site with less available light requires a slower shutter speed to capture enough light. This also requires the drone to fly slower; otherwise the slower shutter speed combined with the drone's movement will cause blurry photos, such as this example. Source: Chinook Landscape Architecture.

From a business/operation standpoint, weather planning is key to keeping a project on schedule and within budget. The data collection portion of a photogrammetry project (the flights) is only the first step. If enough time is not built into the schedule for weather delays, the entire project can be put behind schedule and over budget very quickly.

Ground Control Point Planning. As discussed in the GCP section earlier in this chapter, GCPs are required for the accuracy of a photogrammetry project. GCPs can take a few forms, but they must have known coordinates and they must be visible in the photographs. Because GCPs must be visible in the photographs, advance planning is required. On some projects, there may be multiple points that are already surveyed and are easily visible. On other projects, it may be necessary to create GCPs with survey markers or paint. Although the coordinates can be collected after the flights, the GCPs must be marked prior to the flights so that they are visible during the processing steps (Figure 9.15).

GCP coordinates can be collected through a variety of methods, but it is important to note that the accuracy of the final photogrammetric model will be limited by the accuracy of the GCP coordinate collection. If GCP coordinates are collected with a GPS unit capable of one meter horizontal and vertical accuracy, the maximum accuracy of the photogrammetry model will be one meter.

This accuracy is of course also impacted by the distance to the subject and photograph resolution (resulting in the GSD), the amount of overlap between photographs, and the complexity of the subject(s). GCP coordinates collected with more advanced surveying

Fig. 9-15: GCPs can be permanent survey markers, project specific markers, or they can be as simple as painted marks. Source: Chinook Landscape Architecture and Yeh and Associates.

equipment will result in more accurate photogrammetry models, as long as the photographs are collected properly.

The number of GCPs will vary from project to project, but each processing software gives recommendations based on the type of site, the number of flights, etc. As much as possible, GCPs should be spread fairly evenly across a site, as opposed to a line of GCPs along one side of a project.

RTK Drones, Ground Control Pads

Recently, more drones are being released that include real time kinematic (RTK) GPS receivers. These setups consist of a base receiver, and a rover receiver. The base is set on a tripod, typically on a point with known coordinates. The rover is on the drone. In this type of setup, the base is processing corrections in real time and transmitting those corrections to the rover on the drone. Theoretically, this results in far more accurate geotags on the photographs. At the time of writing this book, these setups still require GCPs, but typically far fewer GCPs per project/area.

In addition to RTK being used on the drones themselves, GCPs can be collected with RTK GPS setups to greatly increase the accuracy of the data. With either setup, the data collected by the RTK receiver can also be post-processed (PPK); this is often done when the base station has a constant connection to GPS satellites, but the connection between the base and rover is intermittent, preventing consistent real-time corrections (Figure 9.16).

Fig. 9-16: Using an RTK GPS setup to collect GCPs. The RTK receiver in the foreground is the base and is set up on a known point. The RTK receiver in the background is the rover and is moved to collect coordinates at each marked GCP. Source: Chinook Landscape Architecture.

Fig. 9-17: Many projects will have unique considerations. This project required detailed traffic control plans which limited the acceptable operational days and times. Source: Chinook Landscape Architecture and HDR, Inc.

Given the number of tech companies working on finding solutions to GCP collection and drones, it's reasonable to expect that simple plug-and-play drone solutions to provide accurate GCPs will be further developed. One example is the Propeller AeroPoints Ground Control pads (https://www.propelleraero.com/aeropoints/). These are self-contained GCP marking and collection units. All AeroPoints must be in place during flights so that they are all visible, and the hardware connects to RTK GPS networks to provide corrected coordinates.

Project Specific Planning. Each project will have unique aspects that are best considered prior to being on-site and flying. These could range from creating traffic closure plans to avoid flying over moving traffic, to contacting local agencies about upcoming flights, to visiting a site in person to determine where the pilot can safely fly from while maintaining VLOS. As part of creating a flight plan, it is important to think through all of these items and plan how each one will be addressed (Figure 9.17).

After preparing the flight plan, it is time to actually fly the drone and collect the data. Even though actually flying the drone is the most visible part of the project and the part that people will associate most with the project, it is just one part of the project. Flight considerations, flight applications, and best practices for flying are covered in Chapters 5–7.

Processing the Data

After collecting the photographs, the work returns to the office. Downloading the photographs onto a computer or server is the first step. As discussed earlier, photogrammetry requires a lot of overlap between the photographs. This results in a very large number of

photographs per project, and of course, a very large amount of virtual storage space. Properly organizing the raw photographs on your computer or server will be key to both processing the data efficiently, and being able to archive project data in order to keep your system running smoothly (Figure 9.18).

Although there are many good ways to organize this data, a few good guidelines or suggestions are:

▶ Organize drone photographs by flight and date.

▶ Organize drone photographs by photogrammetry processing area if possible. For larger projects, it is likely that the overall project will be broken up into smaller projects for the photogrammetry software to process and for everyone to view. Organizing the photographs into these areas will help when processing each area.

▶ Create a standard folder/organization system to be used for all drone photogrammetry projects.

▶ Align the folder/organization system with conventions already used by your organization.

▶ Develop a regular process for archiving project data. Most offices have a regularly scheduled file system backup that backs up the entire file system. Because both the photographs and photogrammetry files are very large, they can significantly slow down this backup. Having a process in place that moves these files onto a long-term storage drive(s) once they are no longer being actively used will avoid this conflict if the long-term storage is not part of the regularly scheduled backup.

Once the photographs have been downloaded from the drone, the photogrammetry processing begins. Although each photogrammetry software package has slightly different options, most follow the same general process. The following steps apply to desktop processing software such as Pix4D Mapper and Agisoft Metashape. There are also simpler online options such as Maps Made Easy; these are more intuitive and require fewer steps. All of these options, including the pros and cons of each one, are discussed later in this chapter. The following process aligns most with the Pix4D Mapper software, although other photogrammetry software uses very similar workflows. The Pix4D Mapper software breaks the processing down into three distinct steps. The steps outlined below are broken down into more detail, and as a result, the step numbers in this book do not correlate to the step numbers in Pix4D Mapper.

Processing Steps

Prior to beginning this processing, it is important to have thought through how many projects will make up the overall project. For a smaller project, it may all be processed together. For larger projects, it may make more sense to split the overall project into smaller projects so that the computer can process the data and everyone can access and view the files. Like with the photographs, it is very helpful to organize the photogrammetry files according to these groupings.

Step 1. The first step of processing involves importing the photographs into the photogrammetry software. Most software will be able to recognize the make and model of the camera, as well as its focal length, sensor size, etc. from the EXIF metadata. EXIF metadata is automatically stored within each photograph and can be accessed by viewing the photograph's properties in File Explorer or a similar application. This is also where the photograph's location data is recorded as GPS coordinates. Most software programs will show the location of the photographs over a low-resolution base map from sources like Digital Globe, Bing Maps, Google Earth, etc. (Figure 9.19).

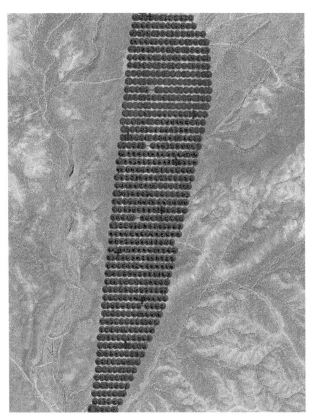

Fig. 9-19: Initial view in Pix4D after beginning a project and importing geotagged photographs (red dots). This screenshot was taken after adding GCPs (blue crosses). Source: North Fruita Desert Trails Master Plan completed by Chinook Landscape Architecture and THK Associates, Inc.

Depending on the software being used, you will need to provide certain inputs, such as flight pattern, desired types of outputs, etc. Once all of this information has been fed into the software, it will create an initial point cloud. The point cloud will display in 3D once processed, and each point will have associated coordinates and a single color. When zoomed away from the point cloud, it will have the appearance of the overall area.

This processing is demanding on computer hardware and can take several hours or more depending on the amount of data and the capabilities of the hardware. The amount of data is not based on the size of the area or how many photographs were taken, but rather how many pixels will need to be processed. For example, a site with 1000 photos taken with a 10 megapixel camera will have far less data to be processed than the same site with 1000 photos taken with a 20 megapixel camera. Each software provides recommended hardware specifications, and general recommendations are discussed later in this chapter.

Step 2. Following the initial processing, manual tie points (MTPs) and GCPs are added to the dataset. In Pix4D Mapper, an MTP is a point that can be visually identified in multiple photographs. A GCP is an MTP that you also have collected accurate coordinates for. Both MTPs and GCPs will provide the software with more data, which it in turn uses to refine the photogrammetry model and add relative accuracy. GCPs also provide the software with known accuracy geolocation data, allowing it to rectify the location of the entire photogrammetry model up to the accuracy of the GCP coordinates (Figure 9.20).

Typically, the GPS data that drones record on photographs is not very accurate; often the horizontal accuracy is within 1–3 meters, and the vertical accuracy can be far less accurate depending on several factors such as signal interference, weather conditions, satellite signal availability, etc. To rectify this and ensure the photogrammetry model is accurately geolocated, GCPs are often collected with higher end GPS equipment or survey equipment.

Fig. 9-20: A GCP marked in Pix4D. The right panel shows the correct coordinates and the error to the initial drone GPS recording. Source: Chinook Landscape Architecture.

Step 3. After creating MTPs and GCPs, and adding the GCP coordinate data, the dataset is re-processed to rectify based on this data. This re-processing takes far less time than the initial processing. In certain cases, the GCP data can be added before the first step, although it is usually very helpful to have the initial point cloud to work from (Figure 9.21).

Step 4. The next step is to create a final point cloud and textured 3D mesh. Most software will offer a variety of options for the point cloud and mesh, ranging from exported file formats, to methods for creating the 3D mesh, to detail of the image texture, to density of the point cloud, to point cloud classification. There is not one set of choices that will be correct for every office or every project. These selections will need to be determined based on the subject of the project, and how the data will be used.

This step is very time consuming, particularly the creation of the point cloud. For larger projects, it is not uncommon for high end computers to take multiple days to complete this step (Figure 9.22).

Step 5. At this point, the point cloud or mesh can be edited to filter out heavy vegetation, normalize planar surfaces, or trim down to a specific area. Most of these edits are made to the point cloud, and then the 3D mesh can be re-created. At this point, point clouds and 3D meshes can also be exported for use in other programs. Depending on the project, this might be the final deliverable, or it might be the base from which design and analysis is conducted.

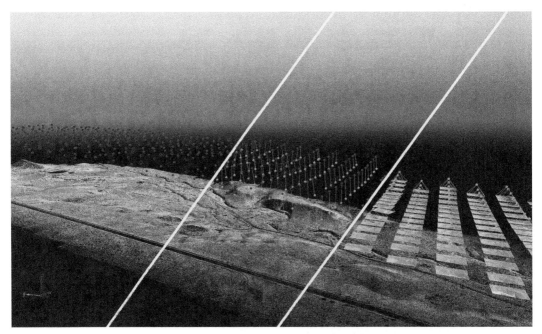

Fig. 9-21: Tryptic showing the initial point cloud and geotagged photo locations (blue dots), corrected photo locations based on GCP input (green dots), and photo angle/direction and thumbnails from corrected locations. Notice the correction from GCP coordinates is mostly vertical as the drone's on-board GPS is fairly accurate horizontally. Source: North Fruita Desert Trails Master Plan completed by Chinook Landscape Architecture and THK Associates, Inc.

Fig. 9-22: Final point cloud viewed in Pix4D. This point cloud contains a little over 80,000,000 points. Source: North Fruita Desert Trails Master Plan completed by Chinook Landscape Architecture and THK Associates, Inc.

Step 6. The final step of the photogrammetry processing is the creation of an orthorectified image, contours, and a digital elevation model (DEM). This step is optional, and how the data for each project will be used should determine what outputs are needed (Figure 9.23).

An orthorectified image is a composition of the entire project area, then rectified to remove perspective distortions. In other words, the orthorectified image is a true plan view image that can be scaled and measured from. The accuracy of this image will lessen at the edges of the project where there are less photographs overlapping the area.

From the 3D photogrammetry model, most software can also create contours and/or a DEM. Contours can typically be created at any set interval (1 foot, 1 meter, 5 feet, etc.), and there are usually additional settings to control things like the resolution of the contours. These settings will vary by project, and often require some trial and error to arrive at the best dataset. DEMs are GEOTIFF files. These are raster images that include elevation data for each pixel within the image. These are most often used in GIS software or Autodesk Civil3D.

Although there are additional options based on the software being used and the specifics of a project, these steps are the basic workflow for creating a photogrammetric model from drone photographs. Below, we will take a closer look at two drone photogrammetry projects that vary greatly in size, subject type, and project objective.

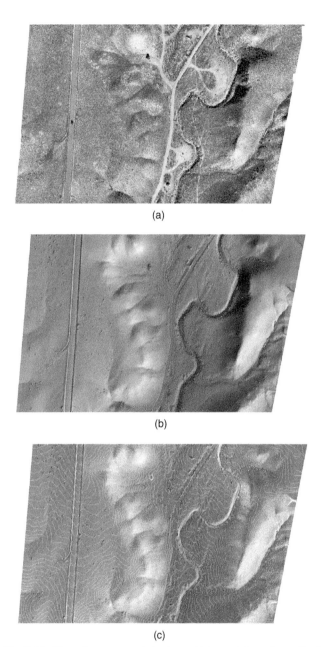

(a)

(b)

(c)

Fig. 9-23: These three screenshots show a high-resolution orthorectified image (a), a DEM (b), and 1 inch contours over the DEM (c) of the same area. Source: North Fruita Desert Trails Master Plan completed by Chinook Landscape Architecture and THK Associates, Inc.

Photogrammetry Project Comparisons

Project #1 – Rural Ranch

This project consisted of collecting data and creating a 3D mesh for a relatively small site of just over 10 acres. The site is mostly bare ground, with some small rock outcroppings and heavier evergreen vegetation along the perimeter of one side. The objective of using drone photogrammetry on the project was to create a base model on which an architectural visualization model of a proposed building could be built. Identifying an approximate building location that offered the best views from particular rooms was also an important factor.

Flight Planning and Operation. In order to capture the photogrammetry photos, three total flights were completed. The first flight was a horizontal grid flight at approximately 200 feet AGL, capturing nadir images. This flight captured photographs with 85% front and side overlap. The next two flights were also horizontal grid flights, but at approximately 120 feet AGL, and with the camera tilted at approximately 45°. The two tilted angle flights were flown at grids perpendicular to each other to capture different angles of vegetation and site features. Because these flights were closer to the subject, they captured photographs with 75% front and side overlap. As mentioned earlier, an important part of this project was to plan for views of the surrounding mountain ranges from rooms within the proposed building. To do so, additional flights were flown to capture a 360° panoramic image at several locations and heights. These 360° images can later be added to the 3D model to simulate the view from different locations within the proposed building (Figure 9.24).

Ground Control Points. The objective of this project was not to create a survey grade model; rather, it was to create a model for visualization purposes and to identify the approximate location of a proposed structure. Later stages in the project will require full surveying and documentation for construction drawing preparation. As such, this drone photogrammetry project could have been completed without any ground control. However, ground control not only accurately geolocates the project (absolute accuracy), but it also increases the accuracy within the project (relative accuracy). For this reason, an RTK GPS setup was used to collect six GCPs across the site. The accuracy of the GCP coordinates was 1 cm horizontally and vertically,

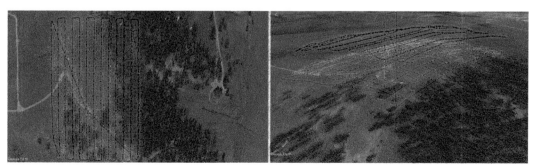

Fig. 9-24: For this project, flight 1 (red path) was a nadir flight flown at 200 feet AGL. Flights 2 and 3 (green and pink paths) were flown at 120 feet AGL and captured oblique imagery. Flights 2 and 3 were flown in grids perpendicular to each other. Source: Chinook Landscape Architecture.

Fig. 9-25: RTK GPS rover set up on a GCP. Source: Chinook Landscape Architecture.

which results in the relative accuracy of the photogrammetry model to be up to 1 cm (depending on image overlap and coverage) (Figure 9.25).

Processing and Delivery. The photogrammetry portion of this project included 1,118 photographs, and they were processed in Pix4D Mapper software. The project was processed on a slightly older workstation with an Intel® Core i7™-4770 3.40 GHZ quad core processor, 32 GB of RAM, and a Nvidia® GeForce™ GTX 970 graphics card with 4 GB of dedicated memory.

The average GSD of the project was 1.69 cm, or 0.67 in. After initial processing, the six GCPs were marked and their coordinates added. The project was then re-optimized which corrected for horizontal and vertical errors in the drone's GPS receiver. As you can see in Figure 9.26, the initial model was fairly close horizontally, but it was corrected quite a bit vertically. The blue symbols each represent a photograph with the original GPS coordinates. The green symbols represent their corrected positions, and the green connecting lines represent the distance of the correction.

Fig. 9-26: Photogrammetry initial point cloud with original photo locations and corrected photo locations. Source: Chinook Landscape Architecture and Viz Graphics.

After re-optimization, the final point cloud and 3D textured mesh were created. Creation of the point cloud took just under 12 hours, classification of the point cloud took about 15 minutes, and creation of the textured 3D mesh took just under 1 hour. The final point cloud contained approximately 90,000,000 points.

After this step, the point cloud was refined to remove the evergreen vegetation, and other artifacts such as the drone pilot's car. The 3D mesh was then regenerated as a "clean" mesh without these features. The "clean" mesh still shows the image of these features, but on a surface that is smoothed to meet the surrounding terrain. See comparison images below (Figure 9.27).

For this project, there is no need for contours, an orthorectified image, or a DEM, so this concluded the photogrammetry portion of the project. From here, the clean mesh was imported into a 3D modeling software, SketchUp in this case. The 360° panorama photographs were stitched together. The SketchUp model and the stitched 360° panoramas were the deliverables for the client to then create the architectural model within.

Project Time. This project required 4 hours of travel, 3 hours of drone operations, and 5 hours of data processing. In all, the project only took 12 hours from start to completion.

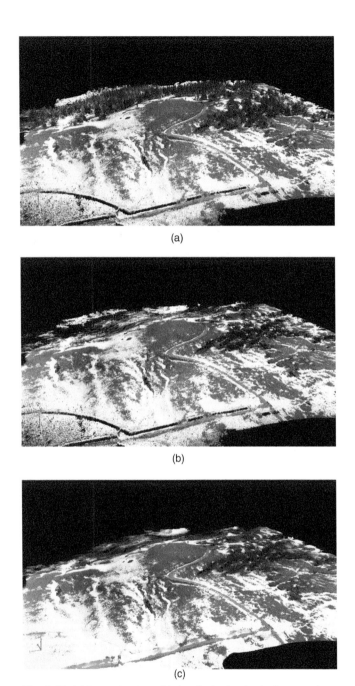

(a)

(b)

(c)

Fig. 9-27: (a) Raw point cloud, (b) classified point cloud with vegetation removed, (c) resulting 3D mesh.
Source: Chinook Landscape Architecture and Viz Graphics.

Project #2 – US-50 Little Blue Canyon

This project consisted of three separate areas along highway US-50 in Southwest Colorado (Figure 9.28). The existing highway has one travel lane in each direction running through a tight canyon with steep rockfaces/cliffs on one side and Blue Creek running along the other. The highway crosses Blue Creek midway through the project site. Because of the tight physical constraints, the highway has sharp curves with poor visibility. The design team has been tasked with developing proposed improvements to smooth the curves, improve visibility around these curves, and generally improve safety. One of the primary constraints is geohazards (rockfalls and rockslides) and our team used drone photogrammetry to capture 3D models of existing rockfaces in the three areas that the geotechnical engineers needed to study.

Flight Planning and Operation. In contrast to the rural ranch project, the subject of this project is vertical, complex, and overhanging terrain. As a result, pre-programmed flight paths were not an option, and nadir imagery by itself would not be useful. Prior to arriving on-site, flight areas and processing areas were determined. Processing areas were defined based on rockface separations (valleys) and the anticipated number of flights per area. For example, it

Fig. 9-28: One of the rockfaces where this project needed a detailed 3D point cloud and mesh for geohazard analysis. Source: Chinook Landscape Architecture.

would make sense to process the first area (rockfaces A, B, and C) as a single project, except that this overall area required 10 flights, which would have created an overly data-heavy photogrammetry project that would be difficult to process and share. This processing certainly is possible on a higher end workstation, although the geotechnical engineer's workflow concentrated on specific areas, and providing a very large dataset of the entire area would not have worked for the engineer's software packages and workflows.

Instead, the area was split into rockfaces A, B, and C. Figure 9.29 shows polygons for rockfaces A, B, and C, as well as the actual flight path lines. The "gap" in the flight paths for rockface C (light blue lines) is where an overhead power line crosses through the site.

Each flight was manually piloted, with an irregular grid pattern that took a series of photographs at different altitudes and angles depending on the subject. The entire project required 16 flights and resulted in approximately 8,500 photographs. This equaled approximately 67 GB of raw photograph data.

Ground Control Points. The photogrammetry model did not need to have survey grade accuracy, although it did need to be accurate enough for geotechnical analysis. For ground control, the engineer collected field marked GCPs with a high-end GPS unit. These GCPs were marked prior to flying. Unfortunately, marking and collecting GCPs on the rockface was not feasible for safety and cost reasons. As a result, GCPs were only collected along the roadway. While not ideal because the GCPs were not spread across the site, they did still provide the accurate geolocation data for GCP rectification.

Fig. 9-29: Flight planning in Google Earth. Source: Chinook Landscape Architecture.

Processing and Delivery. The overall project was split into five projects: rockfaces A, B, C, D, and E. Each project had approximately 1,200–2,200 photographs and were processed in Pix4D Mapper software. The projects were processed on a workstation with an Intel® Core i9™-7900X 3.30 GHZ 10 core processor, 64 GB of RAM, and a Nvidia® GeForce™ GTX 1080 graphics card with 8 GB of dedicated memory.

For this project comparison, consider rockface D in particular. This project had just under 2,200 photographs. Because the drone was flying very close to the subject while capturing photographs, the average GSD was 0.76 cm, or 0.30 in. During processing, four MTPs were created, as well as nine GCPs. The MTPs were placed in areas up and across the rockface in order to provide the software with more information to accurately tie the model together in areas without GCP data.

The point cloud creation took about 11.5 hours and the 3D mesh about 40 minutes (Figure 9.30). The point cloud was not classified. Although this time seems very similar to the rural ranch project at first glance, it is very important to point out the difference in computing power. While the rural ranch project was processed with a four-core processor and 32 GB of RAM, this project used a processor with two and a half times as many cores (10) and twice as much RAM (32 GB). The final point cloud contained approximately 130,000,000 points.

The final deliverable to the geotechnical engineer was the entire photograph library, as well as Pix4D files and point cloud files for each of the rockfaces.

Project Time. This project required 10 hours of travel, 16 hours of drone operations, and 18 hours of data processing. In all, the project took 40 hours from start to completion.

Fig. 9-30: Final point cloud. Zoomed in, the individual points are discernable, but from a distance the point cloud appears solid. Source: Chinook Landscape Architecture and Yeh and Associates.

Photogrammetry Hardware

Photogrammetry generally uses the same hardware as other drone projects discussed in this book. See Chapter 5 for a detailed discussion of the hardware (drones, cameras, and processing computers) used for photogrammetry.

Photogrammetry Software

There are a number of options for photogrammetry processing software. The cheapest and easiest ones are online processing options. Although these will work great for some projects, you have far fewer options and control over the final product. For example, most of them do not offer a way to add GCPs. Most also do not have options to control the quality of the point cloud or mesh, or to edit the point cloud and re-create the mesh. Some online options include Maps Made Easy, DroneDeploy, and Autodesk Recap.

The other type of photogrammetry processing software is desktop software. Similar to other software packages, this is an actual software license that is purchased and downloaded. At the time of writing this book, the most popular desktop photogrammetry processing software packages are Pix4D and Metashape, with RealityCapture gaining a growing following. Pix4D offers multiple packages, but the most widely used is Pix4D Mapper. Pix4D also has cloud processing options that allow users to upload photographs and have Pix4D complete the basic processing. Metashape is made by Agisoft and was previously known as Photoscan. It is offered in a Professional or Standard Edition. RealityCapture is particularly interesting because it can combine laser scans (LiDAR) with photogrammetry. It also claims quicker processing times and less unwanted artifact creation during processing.

Pix4D, Metashape, and RealityCapture are all considered professional grade photogrammetry solutions and offer a wealth of features and content. Although most offices only use one or the other, it is worthwhile to research and use a free trial of each software when determining which one is best for your practice.

In addition to Pix4D Mapper, Metashape, and RealityCapture, there are a number of specialized software packages available for very targeted industries such as mining, construction administration, and security. These are typically much higher priced, and are often packaged with hardware (drone, computers, etc.) and on-site training.

Chapter 10 will further discuss what can be done with the generated point cloud and 3D mesh models. This includes how some of the photogrammetry software can be used to measure the model and point cloud and extract additional data, volumes, and sections.

Working with 3D Models

Photogrammetry, reviewed in Chapter 9, can generate point cloud data and 3D mesh models, allowing for a variety of different uses depending on the project. These 3D cloud and mesh datasets allow for the extraction of site quantities and information useful for design and construction. They can also be used for incredible visualizations. From construction observation and administration to creating visual graphics and renderings, this chapter will provide an overview and specific examples of how to leverage 3D photogrammetric information.

Point Cloud versus 3D Mesh

Let's start with some simple definitions of a point cloud versus a mesh:

Point Cloud. A point cloud, as defined in Chapter 9, is a point in space represented by XYZ coordinates and color values. While a point cloud looks solid when viewed from a distance, zooming in on point cloud data shows the different "points" of the cloud and the gaps between each point. Most point cloud viewers allow you to adjust the size of each individual point; a smaller sized setting shows more of the gaps while increasing the size causes each point to have greater overlap. Adjusting this size is easy to do with most software using a slider (Figure 10.1).

3D Mesh. A mesh is composed of 3D surfaces generated between, and connecting all, the points in a point cloud. 3D software programs interpolate and triangulate distances between the points generating a surface. In total, these surfaces create a solid 3D model; there are no more gaps. 3D meshes offer a different perspective on the data; they tend to look more irregular but dense, displaying the associated texture image (aerial) with greater clarity.

Fig. 10-1: High detail 3D point cloud of an unstable rockface in California. Source: Chinook Landscape Architecture.

3D meshes are typically either triangular irregulated networks (TINs) or quadrangle meshes. As the names suggest, a TIN is made up of triangles (often millions of them), and a quadrangle mesh is made up of four-sided polygons. Most photogrammetric 3D meshes are TINs (Figure 10.2).

Comparison. Each dataset allows for a different way to view the data. The dataset you use will often be based on the software you are using. For example, Maptek Point Studio (a very specialized geotechnical engineering software package) typically imports a point cloud and then creates a 3D mesh that is suited specifically to tasks and tools within Point Studio. SketchUp, Revit, Rhino, 3ds Max, and other mainstream 3D and 2D CAD software have the ability to import point clouds. Using extensions in SketchUp, you can import point clouds (search for Scan Essentials) or mesh data (search for Skimp for SketchUp).

Working with Point Clouds and 3D Meshes

There are three primary approaches to working with 3D data obtained from photogrammetry. Each of these has different user skill requirements and software requirements; the easiest is sharing and accessing 3D views from a computer internet browser, and the most advanced is working with and manipulating the actual 3D geometry in powerful 3D modeling software:

1. Viewing/sharing point clouds and meshes *online* (Figure 10.3).

2. View and process point cloud/3D meshes in *desktop software* (Figure 10.4).

3. Working with and manipulating the 3D mesh (Figure 10.5).

Fig. 10-2: High detail 3D mesh of an unstable rockface in California; this is the same dataset as Fig. 10-1. Source: Chinook Landscape Architecture.

Fig. 10-3: Point clouds, aerials, mesh, and other drone data can be accessed via the cloud through a browser. These links can be sent out to the project team or others for viewing, annotation, and download. Source: Image courtesy of DHM Design.

Fig. 10-4: DroneDeploy, Pix4D, and many other software packages have desktop versions that allow for local processing and manipulation of drone data. Source: Image courtesy of DHM Design.

Viewing and Sharing Online

Most photogrammetric processing software has the option to process the data in the software's computing cloud. The person or company processing the cloud data through the respective software can then send out a simple internet browser link to whomever requires it. This link allows the end users to view the data through the browser and in many instances provides some easy to use and great tools to measure the point cloud or 3D mesh. Most of all, the end user has easy access and simple options to choose from. Some of these tools are:

- ▶ *Navigation*. The users have the option to view and adjust the point cloud or to view the mesh and easily move around the 3D environment in the browser. This requires almost no skill beyond using a mouse and browser.

- ▶ *Measurements*. Many of these browser links come with other tools including the ability to measure distance using a simple "tape measure" like tool. Users can measure vertical and horizontal distance or measure area and volumes of masses, like a dirt mount, an embankment, or rockface.

- ▶ *Sections*. Some software (Pix4D for example) allows users to run a "line" between two points. The cloud-connected browser will then calculate and generate a 2D section cut slice along that line (Figure 10.6).

Fig. 10-5: Drone data can be imported into 3D modeling programs. Crown Mountain Park 3D model, derived from a drone is shown opened in SketchUp 3D modeling software. Source: Image courtesy of DHM Design.

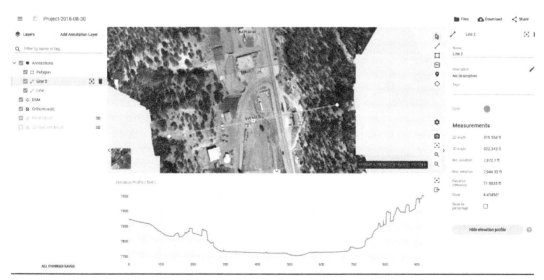

Figure 10.6 Pix4D cloud software allows users to measure distances and volumes, create sections (as seen here), and in general view drone collected data. Source: Image courtesy of DHM Design.

Third-Party Sites

There are many online 3D viewing apps that utilize a browser and allow users to navigate through a model. It requires the photogrammetric data to be processed and then uploaded to one of these sites and most of these sites can only use the 3D mesh (not the point cloud). For example, Sketchfab.com allows you to upload 3D mesh models and either share them publicly (anyone can see the model) or privately (only those with the link or a password can view the model). Once uploaded a simple link can be sent to users to navigate through the model. While not packed with lots of tools, Sketchfab is effective for simply sharing 3D models. The biggest downside to Sketchfab is that model size is limited by your account level (i.e. higher cost account levels allow for larger size models).

A step up from Sketchfab in terms of functionality is the CoLab online platform (Figure 10.7). CoLab is a cloud-based design review and issue tracking platform. Although it was developed specifically for mechanical engineering and manufacturing teams, many infrastructure, architecture, and site design firms have found it to be quite useful. Through CoLab, you can upload full resolution 3D mesh models and invite team members and collaborators. Anyone invited can then make comments and markups, and even create issue tickets that are assigned to other team members. This can be a very effective way for remote or spread out teams to share photogrammetry models and collaborate on issues or observations. This platform can also be extremely useful for reviewing 3D models once you begin modeling proposed features.

Fig. 10-7: CoLab allows for the sharing of 3D models online with collaborative issue tracking and resolution tools. Source: Chinook Landscape Architecture.

Application

Regardless of which software or browser options are used, viewing 3D cloud or mesh information is a great way for project managers and staff to view a site for various reasons. This could range from understanding existing conditions, to tracking information and progress through construction. It provides quick, seamless, and powerful ways to leverage information without requiring a high skill level beyond the company or person processing the data.

Viewing and Processing in the Cloud

Photogrammetry Software

Many photogrammetry software includes its own cloud-based viewing platforms. These cloud-based viewers are very similar to the browser but with more features. Depending on the platform, there are a number of additional tools and options to collaborate between multiple users to assist in project management. In fact, there are powerful construction administration platforms that integrate drone data with management and annotation tools.

Most photogrammetry software needs to be installed on a user's desktop or laptop. In some instances, a software specific browser is installed for project collaboration.

Below are some common features available, some are software specific or dependent:

- ▶ *Project Folders*. Users can access different projects and drone data in an organized file structure.

- ▶ *Drone Data*. Users can access holistic data information: images, video footage, point cloud, aerials, 3D meshes, and other data. It's not just limited to point cloud and 3D mesh information.

- ▶ *Collaboration*. These cloud-based platforms allow end users to collaborate on a project. Users can comment or annotate on datasets, modify and upload useful information, and help work on the project from a remote location or office.

- ▶ *Calculations*. Users can measure distances, volumes, and do cut fill calculations by comparing data. There are more options for the type of data measurements that can be extracted.

Construction Management Viewing and Sharing Software

Some of the more powerful cloud-based drone software like SiteAware, PrecisonHawk, or Trimble Stratus (Figure 10.8) provide specific project tools to better collaborate and track a flown site. One of the most powerful tools is the ability to compare site data between flights, over time. This requires the drone to be flown over pre-recorded and identical flight paths. The flight paths are then re-flown over different intervals: daily, multiple times a week, or less often. As each flight's data is uploaded and processed, the cloud software can process comparisons between the different datasets.

This allows for the tracking of progress on a site such as changes in grade, decrease and increase of material quantities, confirmation of delivery of materials, and other general construction progress. End users will be updated by the software on site changes. This includes the ability to tie these comparisons to Gantt charts, cost estimating spreadsheets,

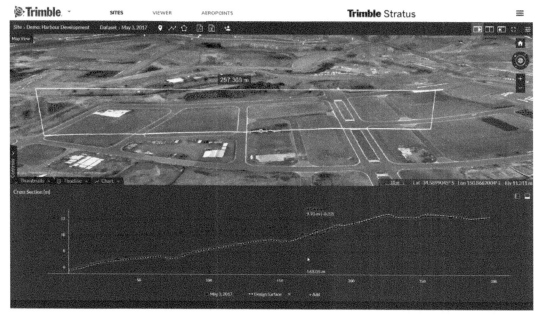

Fig. 10-8: Trimble Stratus is another platform used to process, view, and edit drone captured data. It is a complete project management tool that allows, among other things, users to compare drone data captured over different intervals and time. Source: Trimble YouTube Channel. https://www.youtube.com/watch?v=pRk4QxR3xlU&feature=emb_logo

and even project schedules. Data can be used to evaluate safety conditions, track equipment, and monitor security. In short, it gives unprecedented remote sensing information for a site during construction phases.

The interesting thing about such software is the general ease of use. This does not require a great investment in time for end users. In particular, managers do not need to fly the drone, but are given the ability to easily track a project, ensure costs are kept in check, and ensure progess is being made. The interfaces are simple, not cluttered, with clearly stated options and functions.

Working with and Manipulating the 3D Mesh

There are many reasons to create 3D models from drone photogrammetry data. This can range from infrastructure modeling of roadways, bridges, and channels, to aesthetic rendering for architecture and urban design. For example, point cloud data imported into Civil 3D software provides the opportunity to generate grading files and create roadway or building foundation modeling. Other software like SketchUp, Revit, 3ds Max, AutoCAD, Rhino, and Blender to name a few, each have options for the import of such information, allowing for modifications of a mesh for analysis or modification.

File Format

File format is always important to determine what will best import into any given software. In general, most photogrammetry software can produce or export an FBX, OBJ, and/or Collada format file. These are a solid range of formats that can be imported by most 3D modeling software. As important, these formats can be imported into decimation software referenced below.

Point Cloud to Mesh

While it's possible to use point cloud data in all the software listed above, to modify the drone data typically requires the information to be a 3D mesh. Most of the software mentioned above can import point cloud data and most of them can generate 3D mesh from the point cloud (Figure 10.9). However, this is not always the most ideal way to work with a point cloud generated 3D mesh.

It's important to have a clean, manageable mesh to generate 3D renderings from (Figure 10.10). Being cleaned up refers to a classified point cloud (see Chapter 9), which results in a site model where vegetation and other vertical objects are placed on layers and can be turned off. This provides a clean base model.

Mesh Decimation

However, before importing the classified mesh into software, it's important to decimate the mesh (i.e. reduce the total number of surfaces that compose the dataset). As mentioned earlier, most photogrammetry meshes are TINs, so another way to think of decimation is reducing the number of triangles. This makes the base mesh easier to work with as it's a much smaller file. It allows for better performance and makes modifications/manipulations easier. Software like Rhino can decimate these types of meshes with ease. However, Rhino is limited in exporting

Fig. 10-9: Clear Creek Canyon pedestrian bridge 3D model integrated into the 3D mesh derived from drone. The drone captured photogrammetry was done with a high level of detail in mind with the goal of generating a highly detailed, photo real quality mesh. Source: Chinook Landscape Architecture.

Fig. 10-10: Clear Creek Canyon pedestrian bridge 3D model is integrated into the 3D mesh derived from drone photogrammetry. Source: Chinook Landscape Architecture.

successfully to other software. Other software packages need a little help. In particular Revit, SketchUp, and to a lesser extent, 3ds Max. Using third party decimation software prior to import becomes important and these apps are readily available.

For SketchUp, one option is the Skimp extension (Figure 10.11). This allows for the import of the mesh and can decimate or reduce the mesh size by 90% or more. Decimation makes it ideal to work in SketchUp and other software. Others options are Atangeo Balancer and Meshlab. All of these programs are fairly simple to use. Load the mesh data from the photogrammetry software, usually this would be an FBX or OBJ format as mentioned above. Then, using a simple slider or value entry indicate the amount of decimation. As much as 99% reduction if needed, although the model will likely lose too much detail at 99% decimation. For SketchUp users, it's recommended to decimate a mesh in order to arrive at a final mesh with fewer than 500,000 faces, or triangles. The correct decimation percentage will vary based on the project, and it is important to keep an eye on where details are being lost or retained at different decimation percentages. This will maintain a solid level of performance.

Mesh Manipulation and Proposed Features

Modeling design concepts and overlays in 3D using an existing mesh captured from a drone provides an ideal base for rendering and visualizations (Figure 10.12). Once the mesh is imported into programs like SketchUp, Rhino, or Revit, it becomes simple to render the model using apps like Lumion, Enscape, or V-Ray (just to name a very few).

Examples of these types of modeling are mentioned in Chapter 8 when matching drone photos to 3D modeling. In this instance, design concepts like roads, bridges, urban plazas,

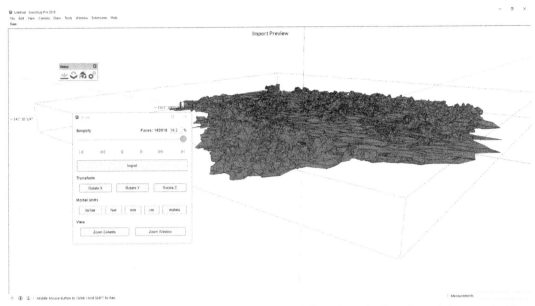

Fig. 10-11: The SketchUp extension SKIMP allows for the import of 3D mesh created from drone photogrammetry. The extension allows for the decimation of the mesh size, making it easier to import and work with. Source: Image by Daniel Tal.

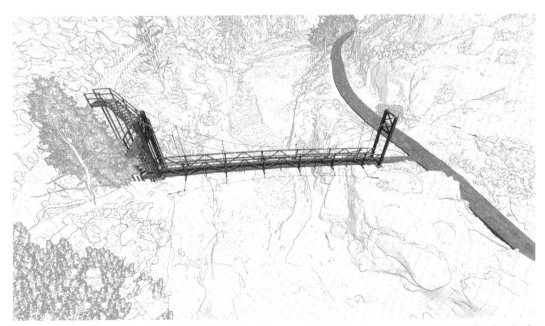

Fig. 10-12: The bridge and trail, modeled in SketchUp, is incorporated into the drone mesh, then imported into Lumion for final rendering. White areas represent the drone data 3D mesh, while the colored bridge and trail represent the proposed features modeled in SketchUp. Source: Chinook Landscape Architecture.

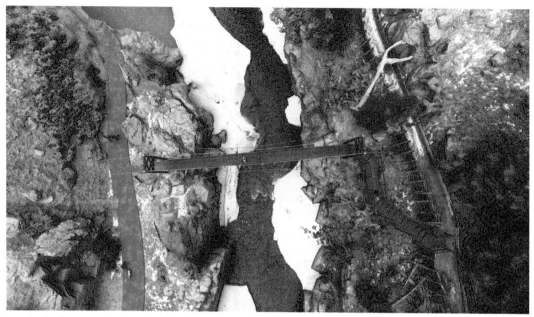

Fig. 10-13: The combined power of 3D modeling and drone photogrammetry allows for a compelling approach to any type of design project. Source: Chinook Landscape Architecture.

and architecture are set into the drone base model. If the model was classified, trees and other vertical objects that were removed can simply be re-added with higher quality (not mesh-model, blobby looking) vegetation, cars, people, and other exterior objects and entourage. These types of results are usually harder to come by without drone photogrammetry data. The inclusion of the drone base file, a natural extraction from most drone data, saves time and provides accurate, high-resolution textures conveying a stronger sense of realism.

In addition to modeling proposed features onto the drone photogrammetry mesh base, grading changes can be modeled. Once imported into a 3D modeling software such as SketchUp, the photogrammetry mesh is 3D geometry, just like any other 3D geometry. Cuts, fills, and other grading changes can be modeled directly into the 3D mesh.

Environmental modeling software Lumion and 3ds Max are ideal to use with this type of data. Their large libraries of vegetation, people, and objects allow for a complete dress up of the model to produce conceptual renderings ideal for presentation and public engagement (Figure 10.13).

For example, for this bridge concept model, the classified mesh is first imported into SketchUp. The 3D bridge and trail are then modeled in SketchUp within the site location. Site plan modifications can be included as needed. The model is then imported into Lumion. Here, shrubs and trees are added over the aerial in appropriate locations. The higher quality vegetation library allows for wide variation to create accurate context. Adding people, cars, and other objects using the Lumion and SketchUp 3D Warehouse libraries completes the model. Also in Lumion, the sun, weather, and other filters and effects can be added to produce renderings or animations.

CHAPTER
11

The Future of UAVs

Drone technology has disrupted how we collect and view data in our surroundings. What once took weeks of field collection can now be collected in a matter of days, often hours. Areas once considered unreachable, or at least uncollectable from a data perspective, are now thought of as no problem. The precision of the data collected via drones makes us question how we traditionally collect data and what that precision means. On top of this, technologies traditionally used in a lab or very small scale settings are finding new uses in landscape scale projects because of drones.

Although this book focuses on site and physical data in architecture, engineering, and construction industries, the application of drone technology already reaches far beyond these. Within these industries, drones are also being used with LiDAR sensors, thermal sensors, infrared/multispectral sensors, and NDVI sensors (Figure 11.1).

Within other industries, drones are being used in a plethora of new ways. Drones have been used to uncover hidden archaeological sites with LiDAR and ground penetrating radar. Entire software packages have been developed to use drones for construction administration and monitor construction progress matched up with BIM design drawings. Full turn-key operations for automated drone site surveillance have been developed for security operations. Drones are being used to monitor ice shelf changes and movements. Drones have helped to track wildlife poachers and protect endangered species.

The one commonality between all of these is that the drone is simply a platform to place data collection sensors in hard-to-reach vantage points much more cheaply and quickly than traditional platforms such as airplanes, satellites, or ground vehicles. Below are a few examples of where drone technology is already progressing:

Fig. 11-1: Thermal drone imagery (dataset courtesy of Pix4D). Source: Jon Altschuld.

NDVI Sensors and Plant Identification. The ability to use infrared, Normalized Difference Vegetation Index (NDVI), and other sensors to measure and quantify plant and ecosystem information is still in its infancy but will quickly grow into its own market segment for drones. The basic premise is that healthy vegetation reflects more, or a better quality of, light compared to less healthy flora. This allows for the measurement of evapotranspiration. This is most often measured through the NDVI, which measures the level of green color in leaves to assess plant health and level of moisture or water needs. NDVI, specifically on drones, has been pioneered largely by the agricultural industry and is already widely used for crop production and management (Figure 11.2).

For some quick information go to:

https://gisgeography.com/ndvi-normalized-difference-vegetation-index/

https://support.dronedeploy.com/docs/understanding-ndvi-data

Gas Leak Detection. Custom drones can be outfitted with sensors to detect gas leaks. This is ideal for inspection of gas and power facilities, pipelines that traverse hundreds of miles, mines, and more. There are already specific drone models to do this type of work allowing for quicker and safer detection. These drones are highly specialized and prices typically start around $40,000–50,000. See the following examples:

https://viper-drones.com/systems/dji-m600-pro-optical-gas-imaging-gas-leak-detection-system/

http://ulcrobotics.com/services/uav-gas-leak-detection/

Fig. 11-2: NDVI image of a sequoia tree (https://www.youtube.com/watch?v=XF306Hp6Q4I).

Ecopia Geospatial. Ecopia (https://www.ecopiatech.com/) is a Canadian start up that provides AI driven remote sensing products. Using advanced algorithms Ecopia is able to generate building footprints and building heights from aerial imagery. One of the products/services Ecopia offers is the ability to generate line work and edges from aerials to create a 2D line survey of a location. Ecopia software can generate curbs, gutters, walks, walls, and building footprints, and identify and annotate plant locations, parking lots, sports fields, landscape and forests, and more. They provide this information as a DWG, JSON, shapefile, and other formats. Customers can fly a drone, collect the aerial data, and send this data to Ecopia who will generate the desired line work (Figure 11.3).

Automated "Drone Boxes." An up and coming technology are on-site drone storage boxes with computer hardware and automated flight tools. These have many functions. For example, these can be placed at a location while a drone operator hundreds or thousands of miles away can access the drone box and drone stored inside. The operator can order the drone to fly and scan a location and return to the box. The box contains the hardware needed to process data and share telemetry and information with the user. Although this is currently beyond what is allowed under FAA rules, the FAA rules are quickly evolving and rules/regulations vary by country.

Fig. 11-3: Ecopia Tech uses computer AI to identify and segment a high-resolution aerial into an accurate CAD drawing. Source: Image by Daniel Tal.

Some example links:
https://www.airoboticsdrones.com/
https://percepto.co/solutions/

Drone Delivery Systems. One of the most common and well publicized drone innovations is the use of drones to deliver packages of various sizes to locations. This will take time to develop as countries figure out how to integrate such devices into airspace, not to mention the current limitations on flight times and battery life. However, both Google and Amazon have secured certification from the FAA to operate delivery drones within US airspace.

Sentries and Security. From small single-family home security drones, to perimeter drones, to police observation and reconnaissance drones, security has been part of the drone industry from the very start of unmanned flights. Among other things, this has sparked an interesting debate for police departments across the US as to whether they should implement drones. Some have opted out, while others like Memphis, have a drone fleet that help patrol the streets. Many police and rescue organizations have developed drone fleets to assist with search and rescue operations.

This is a growing industry that will be part of the larger community, facility, and home security field in the coming years. See the following example:
https://sunflower-labs.com/

Passenger Drones. From UBER to Boeing, passenger drones might make their debut before driverless cars become a full-scale reality. As of 2019, they are already being tested in

Australia by Uber. Passenger drones will offer the ability for short hops. As airspace rules are being reviewed for greater inclusion of commercial drones, passenger drones or drone taxis will lift people from airports to local destinations, similar to other mass transit. The reduced costs and requirements for large infrastructure can potentially launch this part of the industry into overdrive quickly.

Below are some example links:

https://passengerdrone.com/

http://www.ehang.com/ehang184/

https://www.uber.com/us/en/elevate/

Final Word and Looking Ahead

Drones are an emergent, disruptive technology. Market growth and new technologies will define the uses of this versatile technology. If you read this book, follow the steps, and become part of the expanding drone technology sector, then you are part of the future and wherever that might go.

Some quick suggestions for readers to keep an eye on the future:

1. Keep your license current every two years. You never know what having a license might mean in the future for work and, even better, the chance to play with new and cutting-edge technology.

2. Make sure to explore different software and hardware. Don't be married to a single type of drone manufacturer or software. Things are always changing and there are new players in both markets.

3. Share knowledge and information with other drone users. This is the best way to learn about new tech and current trends.

4. Have fun. You get to fly a small, portable plane. It's the stuff of science fiction.

Fig. 11-4: Jon Altschuld flying a Phantom 4 Pro. Source: Jon Altschuld.

As the drone industry moves forward, we encourage you to explore other uses of drones within your industry, whatever that may be. Look for opportunities to combine drones with other technologies, sensors, software, and experts to arrive at a solution for whatever unique problem your project faces.

Index